From the Nation's #1 Educational Publisher K-12

Grade 1

Enrichment Math

Challenging and Fun Activities
Patterns • Fractions
Time & Money • Problem Solving

Table of Contents

Counting 3	Numbers to 100 21
Counting to 10 4	Number Order 22
Fact Families 5	Skip-Counting 23
Number Patterns 6	Progress Check 24
Problem Solving 7	Problem Solving 25
Adding Three Numbers 8	Time—Hour 26
Subtraction Facts 9	Time—Half Hour 27
Subtraction Patterns 10	Review 28
Addition and Subtraction 11	Pennies and Nickels 29
Fact Families 12	Pennies, Nickels, and Dimes 30
Facts to 10 13	Quarters 31
Problem Solving 14	Problem Solving 32
Addition and Subtraction 15	Order to 100 33
Numbers to 39 16	Centimeters 34
Numbers to 59 17	Problem Solving 35
Greater and Less 18	Inch and Foot 36
Problem Solving 19	Problem Solving 37
Numbers to 79 20	Progress Check 38

Table of Contents (continued)

Facts to 11 39	Facts to 15 53
Problem Solving 40	Problem Solving 54
More or Less 41	Problem Solving 55
Problem Solving 42	Tens 56
Facts to 12 43	Tens and Ones 57
Problem Solving 44	Adding Ones and Tens 58
Patterns 45	Review 59
One Half .46	Progress Check 60
One Fourth 47	Problem Solving 61
One Third 48	Subtracting Ones and Tens 62
Problem Solving 49	Progress Check 63
Progress Check 50	Estimating 64
Facts to 13 51	Answer Key 65
Facts to 14 52	

Credits:
McGraw-Hill Consumer Products Editorial/Production Team
Vincent F. Douglas, B.S. and M. Ed.
Tracy R. Paulus
Jennifer Blashkiw Pawley

Design Studio
Cover: Beachcomber Studio
Interior: Color Associates

Warner Bros. Worldwide Publishing Editorial/Production Team
Michael Harkavy   Charles Carney
Paula Allen   Allen Helbig
Victoria Selover   Holly Schroeder

Illustrators
Cover & Interior: Animated Arts!™

McGraw-Hill Consumer Products
A Division of The McGraw-Hill Companies

Copyright © 2000 McGraw-Hill Consumer Products.
Published by McGraw-Hill Learning Materials, an imprint of McGraw-Hill Consumer Products.

Printed in the United States of America. All rights reserved. Except as permitted under the United States Copyright Act, no part of this publication may be reproduced or distributed in any form or by any means, or stored in a database or retrieval system, without prior written permission from the publisher.

The McGraw-Hill Junior Academic Series and logo are trademarks of The McGraw-Hill Companies © 2000.

ANIMANIACS, characters, names, and all related indicia are trademarks of Warner Bros. © 2000.

Send all inquiries to:
McGraw-Hill Consumer Products
250 Old Wilson Bridge Road
Worthington, Ohio 43085

1-57768-281-5

2 3 4 5 6 7 8 9 10 QPD 04 03 02 01 00 99

 NAME Matthew Ming

COUNTING

Look at the picture. Write the addition sentence. Find the sum.

1. How many 🐰 and 🐰 in all? 4 + 2 = 6

2. How many 🐵 and 🐵 in all? 6 + 2 = 8

3. How many 🦛 and 🦛 in all? 4 + 1 = 5

4. How many 🐻 and 🐻 in all? 1 + 8 = 4

5. How many 🦒 and 🐰 in all? 3 + 1 = 4

3

Counting to 10

The Warner Bros. Animation Department has 10 floors. Write the number.

1. Slappy is on 4.
   How many more floors to the top? __6__

2. Pinky is on 2.
   How many more floors to the top? __8__

3. You are on 6.
   How many more floors to the top? __4__

4. Skippy is on 1.
   How many more floors to the top? __9__

5. The Brain is on 3.
   How many more floors to the top? __7__

6. You are on 9.
   How many more floors to the top? __1__

Fact Families

Match the fact family numbers to the domino model.

1 2 1

0 3 3

1 2 3

2 2 4

2 2 0

5 0 5

3 2 5

1 5 4

5

Number Patterns

Continue the pattern.

1. 6, 3, 1, 6, 3, 1, __6__, __3__, __1__   88%
2. 2, 2, 4, 2, 2, 4, __2__, __2__, __4__
3. 8, 0, 8, 0, 8, 0, __8__, __0__, __8__, __0__
4. 2, 4, 6, 2, 4, 6, __2__, __4__, __6__
5. 1, 2, 1, 3, 1, 4, __5__, __6__, __7__, __8__    Look carefully!
   1, 5, 1, 6

Make up your own number patterns.

6. __8__, __0__, __0__, __8__, __0__, __0__, __8__
7. __2__, __5__, __8__, __4__, __2__, __5__, __8__
8. __4__, __8__, __4__, __8__, __4__, __8__, __4__
9. __0__, __1__, __2__, __3__, __0__, __1__, __2__

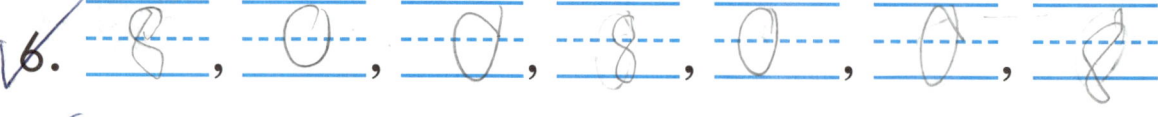

Problem Solving

Use the chart to answer the questions.
Each picture on the chart stands for 1 child.

HOW WE GO TO SCHOOL

walk	👦	👦	👧	👦	
ride in car	👦	👧	👧		
ride in bus	👦	👧	👦	👦	👧

1. How many children walk or ride in a car? _7_   4+3=7

2. How many children walk or ride the bus? _9_   4+5=9

3. How many children ride in a car or bus? _8_   3+5=8

ADDING THREE NUMBERS

Help Mr. Plotz get to his trophy. Choose the right road.

1. 2, 4, 7, 1 / 3, 5 → 9

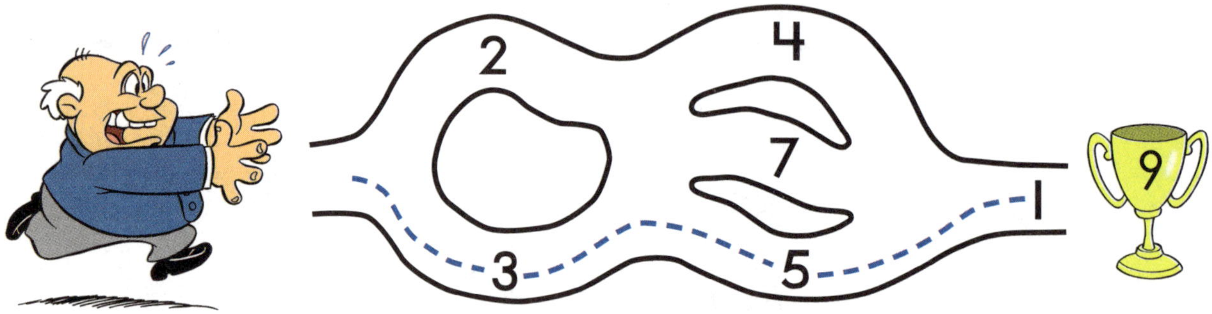

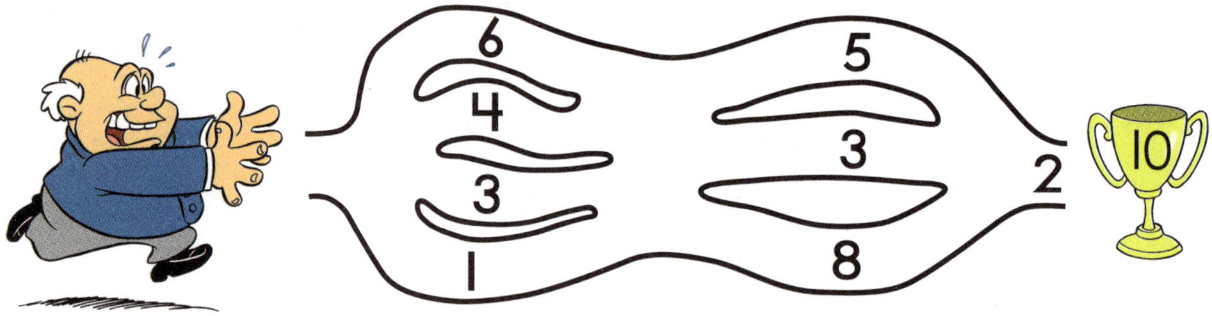

2. 6, 4, 3, 1 / 5, 3, 2, 8 → 10

3. 3, 4, 2 / 6, 2 / 3 → 9

4. 7, 4 / 6, 1, 5 / 2, 0, 3 → 10

8

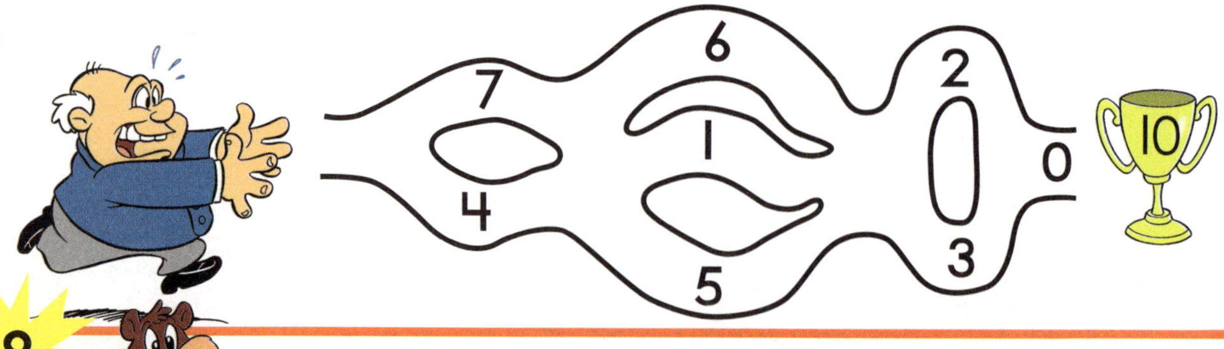

Subtraction Facts

Follow the directions to find the numbers.   + moves ↑   − moves ↓

	Start On	Move	Number
1.	4	−1	
2.	1	−5	
3.	−6	+4	
4.	0	+3	
5.	−2	−2	
6.	−3	+6	
7.	2	−5	
8.	−5	+2	
9.	6	−5	
10.	1	+3	
11.	6	−3	

Name

Subtraction Patterns

Look at the shapes and numbers.

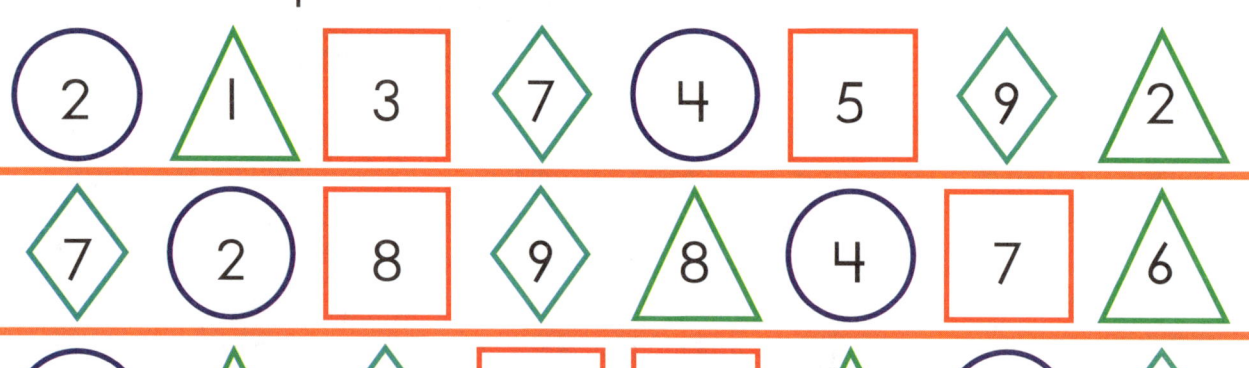

Write the number on each shape in each row. See if the numbers make a pattern. Circle yes or no.

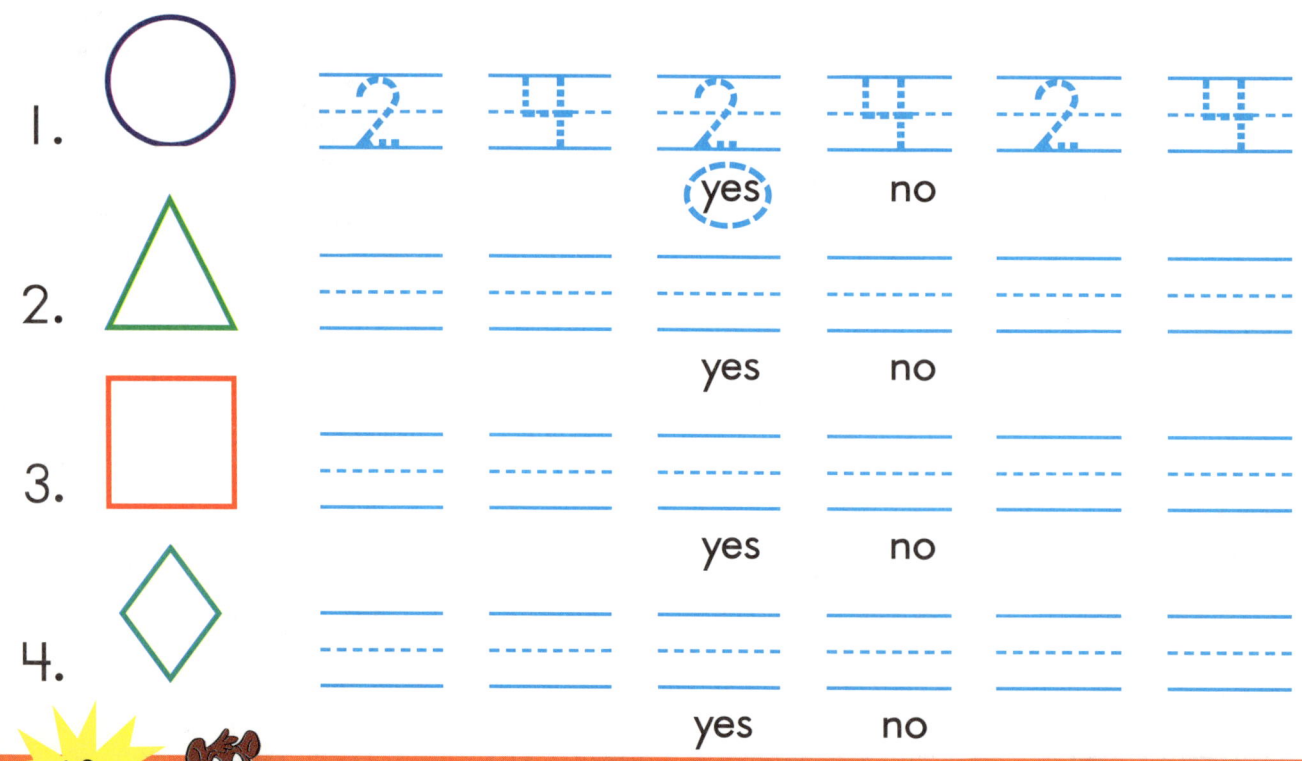

NAME

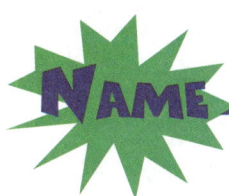

Addition and Subtraction

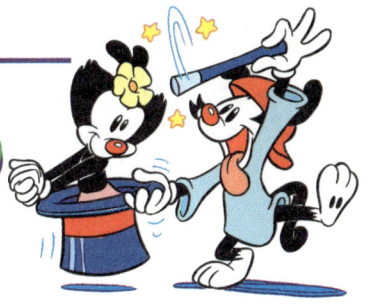

Write 0, 1, or 2 on Wakko's wacky sack. Add or subtract.

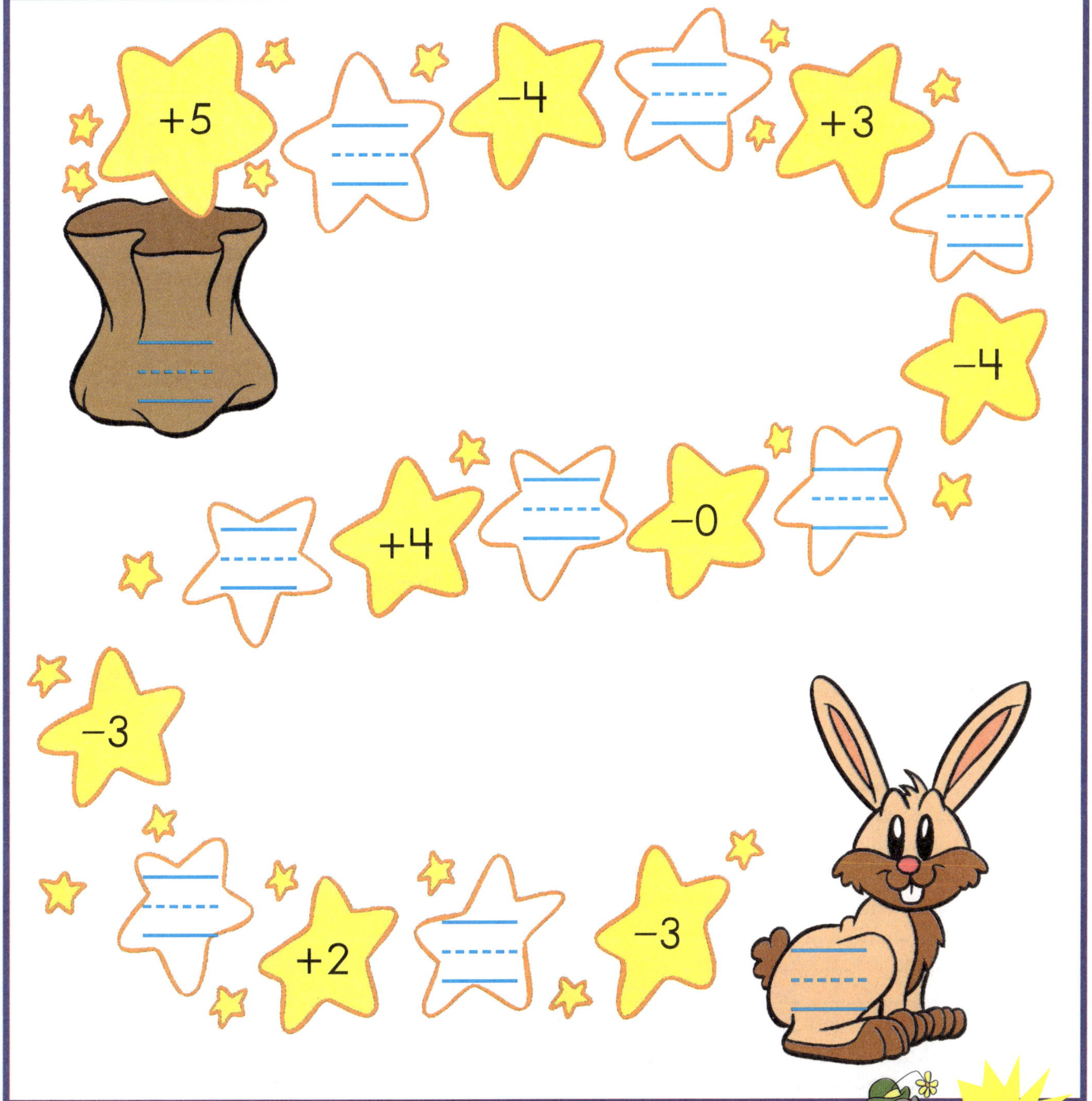

NAME ____________________

HOME
SWEET
HOME

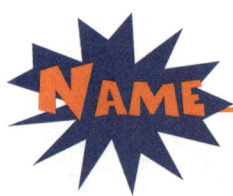

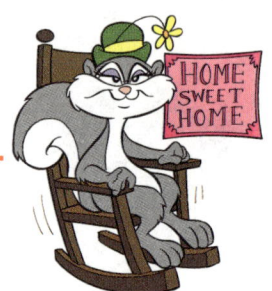

FACT FAMILIES

Use the numbers to write a fact family.

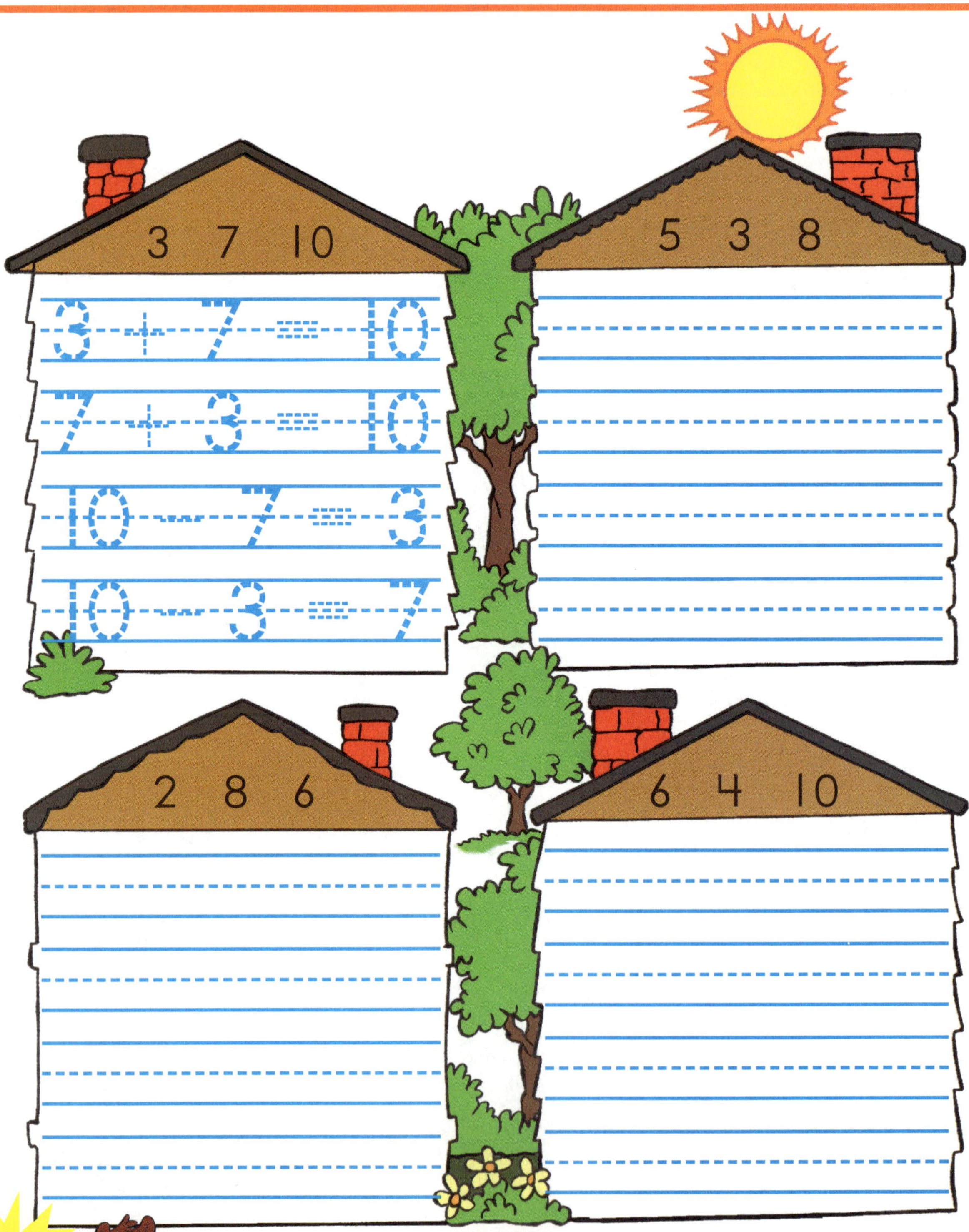
3 7 10
3 + 7 = 10
7 + 3 = 10
10 − 7 = 3
10 − 3 = 7
5 3 8
2 8 6
6 4 10

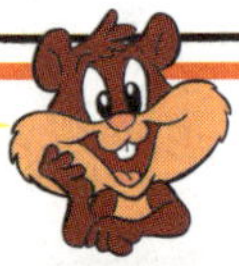

FACTS TO 10

This is a subtraction machine. It subtracts the number in the window.

The number goes in here. →

← The answer comes out here.

The machine subtracts.

Write the answer that comes out of each machine.

1. 10 − 2 = 8   9 − 3 = ___   10 − 7 = ___
2. 9 − 2 = ___   10 − 5 = ___   8 − 2 = ___
3. 10 − 4 = ___   10 − 1 = ___   9 − 5 = ___
4. 7 − 4 = ___   8 − 3 = ___   10 − 8 = ___

Problem Solving

Read. Solve.

1. Mindy made 6 🥪.
   She had 2 🥪 left.
   How many 🥪 did Mindy eat?

   4

2. 8 🐜 were working.
   Later, only 5 🐜 were working.
   How many 🐜 stopped working?

3. Kelley packed 9 🍎.
   Later, 3 🍎 were left.
   How many 🍎 had been eaten?

4. Rich bought 7 🧺.
   He returned 4 🧺.
   How many 🧺 did Rich keep?

5. Buttons saw 10 🥧 in the park.
   Only 3 🥧 were left at the end of the day.
   How many 🥧 were eaten?

Addition and Subtraction

Write the missing signs.

1. $2 + 2 = 8 \bigcirc 4$
2. $5 + 4 = 3 \bigcirc 6$
3. $2 + 6 = 10 \bigcirc 2$
4. $7 + 1 = 9 \bigcirc 1$
5. $7 - 5 = 3 \bigcirc 1$
6. $10 - 7 = 2 \bigcirc 1$
7. $7 - 4 = 9 \bigcirc 6$
8. $8 - 3 = 3 \bigcirc 2$
9. $10 - 2 = 5 \bigcirc 3$
10. $9 - 6 = 5 \bigcirc 2$

NUMBERS TO 39

The inside circle shows tens. The outside circle shows ones. Color to show how many points.

1. 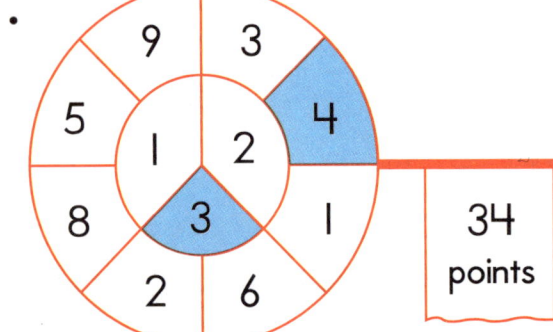 34 points

2. 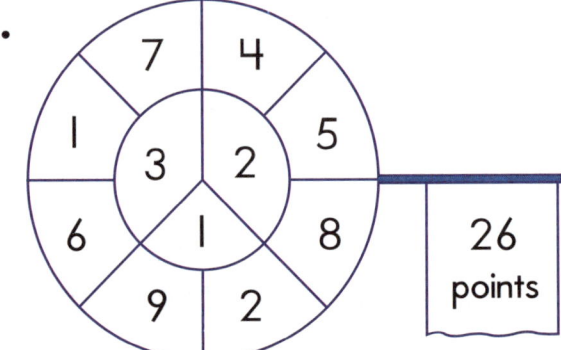 26 points

3. 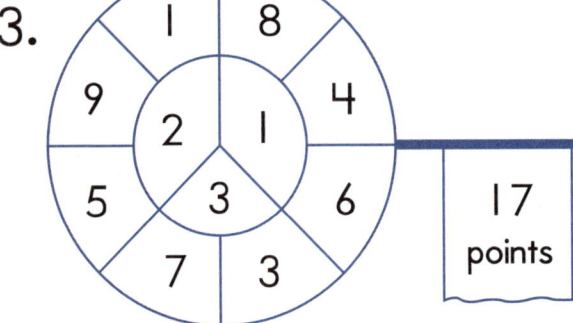 17 points

4. 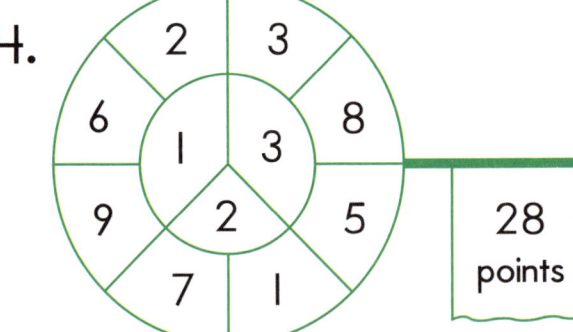 28 points

Write how many points.

5. 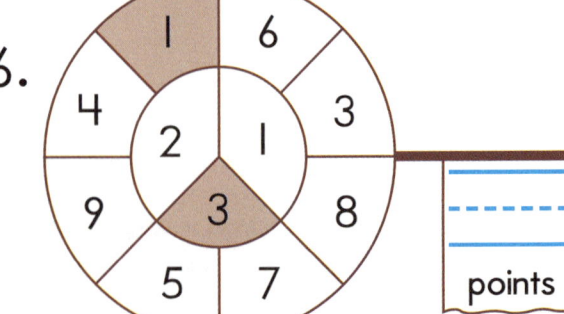 points

6. points

Numbers to 59

The class puts on a play. The chart shows how many tickets each student sold.

 = 10 tickets    = 1 ticket

NUMBER OF TICKETS SOLD

|  3  2  |  5  6  |
|  5  3  |  4  8  |

Use the chart to answer the questions.

1. How many did  sell? __53__

2. How many did  sell? _____

3. Who sold 32?

4. Who sold the most?

5. How many did  sell? _____

Greater and Less

Name _____

Draw a line from the riddle to the correct number.

1. A number greater than 44.
   A number less than 50.
   Count by fives to find it.

2. A number less than 27.
   A number greater than 24.
   Count by twos to find it.

3. A number greater than 20.
   A number less than 40.
   You say it when you count by fives and twos.

4. A number less than 55.
   A number greater than 22.
   A number with two digits that are the same. Count by twos to find it.

26

30

44

45

NAME _____

Problem Solving

Read. Circle the correct question. Then write the correct number sentence.

1. Hello Nurse has 5 beach umbrellas. Wakko gave her 4 more.

   (How many umbrellas does Hello Nurse have in all?)

   How many umbrellas does Hello Nurse have left?

   5 + 4 = 9

2. Wakko has 4 pairs of sunglasses. Hello Nurse has 3 pairs of sunglasses.

   How many pairs of sunglasses do they have in all?

   How many pairs of sunglasses do they have left?

3. Wakko had 10 beach balls. He gave 6 beach balls away.

   How many beach balls does Wakko have in all?

   How many beach balls does Wakko have left?

 **Name** _____

Numbers to 79

 "The cakes show that I am 73 years old."

Write the age the cakes show.

1.

   45

2.

   _____

3.

   _____

4.

   _____

Draw candles on the cakes. Write the age the cakes show.

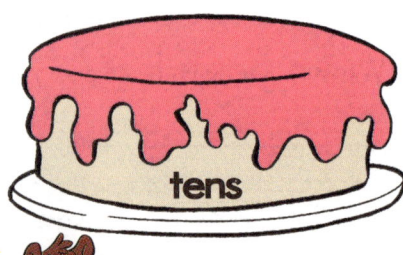    _____

NUMBERS TO 100

Hello Nurse puts numbers on race cars.
She picks 2 digits for each car.

I could make the number 78.
I could also make the number 87.

Write the two numbers Hello Nurse could make.

1.

2.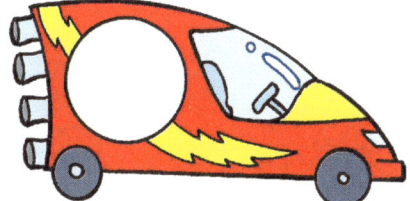

3.

4.

21

Number Order

Connect the numbers in numerical order. Start at GO. End at STOP.

GO
30	35	34	35	44	45	46
31	32	33	36	43	42	47
33	34	38	37	42	49	48
35	36	39	40	41	50	51

STOP

GO
79	80	85	86	95	97	98
80	83	84	87	94	95	96
81	82	89	88	93	98	97
82	89	90	91	92	99	100

STOP

22

Skip-Counting

Mindy and 6 of her friends live on the same street.

Solve.

1. Mindy went to Rita's house.
   Then she walked back 2 houses.
   Where is she now?

   Yakko's house

2. Pinky went to Yakko's house.
   Then he walked on 3 houses.
   Where is he now?

3. Mindy went to Ralph's house.
   Then she walked on 3 houses.
   Where is she now?

4. Dot went to Brain's house.
   Then she walked back 5 houses.
   Where is she now?

 _____

PROGRESS CHECK

5    2    3

Write how many .

1.     12

2.    _____

3.     _____

4.     _____

5.    _____

24

NAME ______________________________

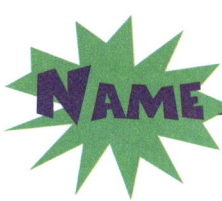

PROBLEM SOLVING

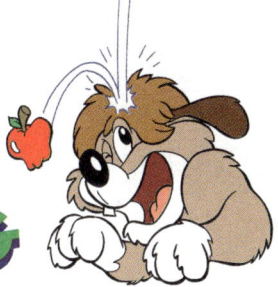

Find how many. Color to complete the graph.

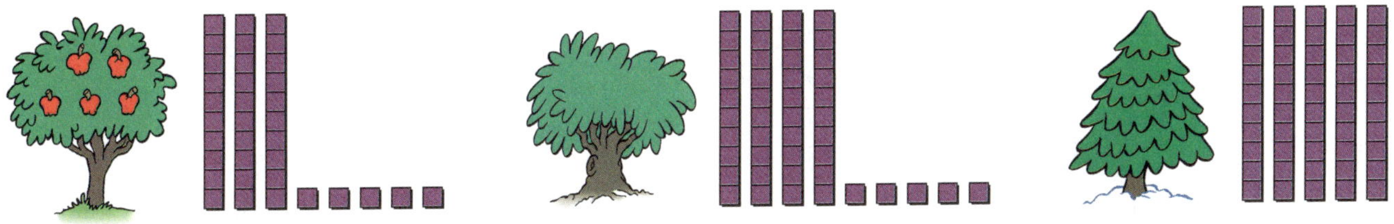

TREES PLANTED

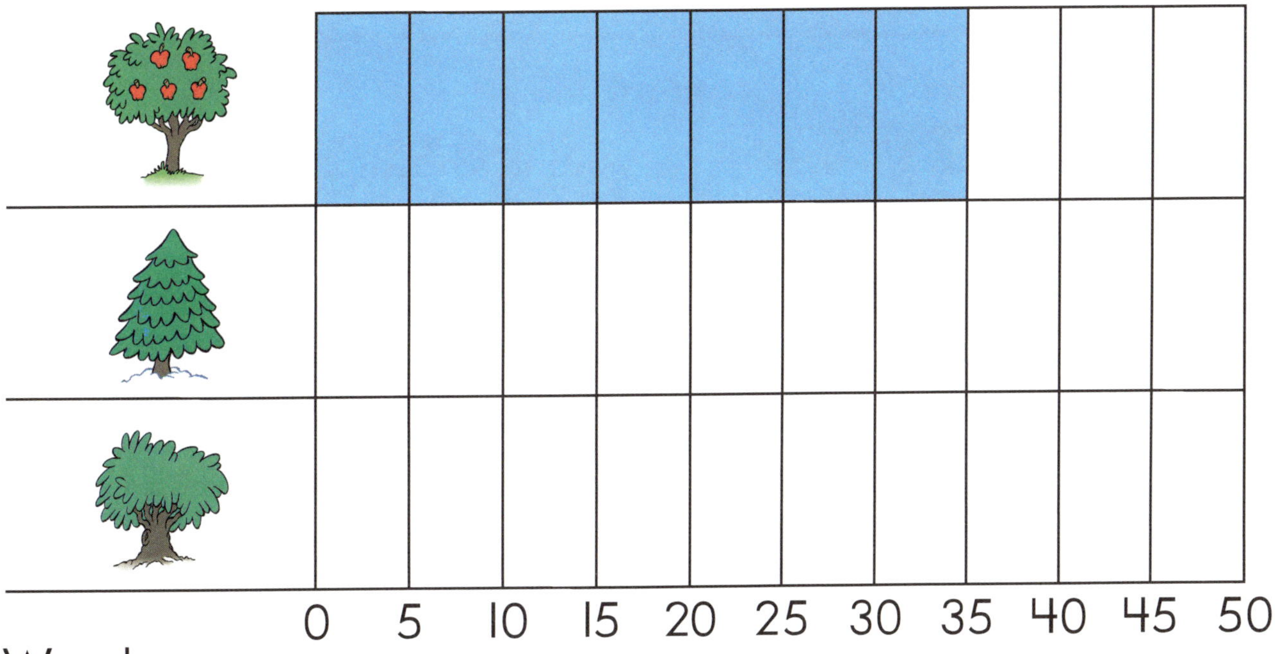

Write how many.

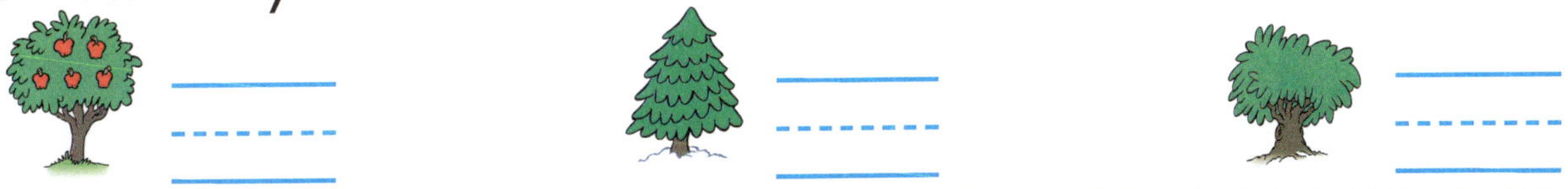

Circle the tree that was planted the most.

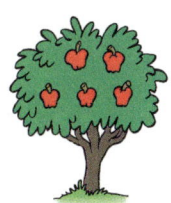

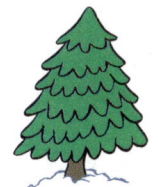

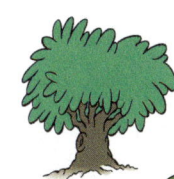

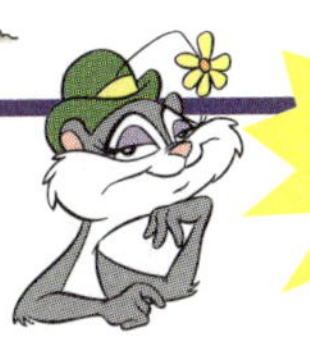

Time—Hour

Both clocks show 8 o'clock.

morning

night

Draw the hour hand to show the time.

Circle: morning  night

1.    6 o'clock

2.    7 o'clock

3.    12 o'clock

4.    9 o'clock

NAME _____

Time—Half Hour

Mr. Tock fixes slow clocks.

Show the correct time.

1.  1 hour slow  ⁻3⁻ o'clock

2.  4 hours slow  _____ o'clock

3.  2 hours slow  _____ o'clock

4.  6 and one half hours slow 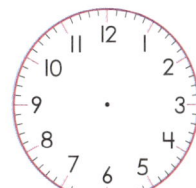 _____ minutes after _____ o'clock

27

Review

Look at each row. Find the pattern. Show the times that come next.

1. [clock 7:30] [clock 9:00] [clock 10:30] [clock 12:00]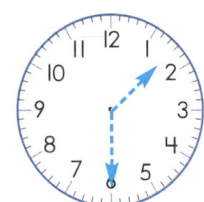

2. | 6:00 | 9:30 | 1:00 |  |  |

3. [clock 1:00] [clock 4:30] [clock 7:30] [blank clock] [blank clock]

4. | 11:30 | 5:00 | 10:30 | : | : |

Make up your own patterns.

5. [blank clock] [blank clock] [blank clock] [blank clock] [blank clock]

6. | : | : | : | : | : |

28

Pennies and Nickels

Put an X on the coins to match the amount.

1. kazoos

2. harmonicas

3. cymbals

4. drums

5. horns

6. banjos

Pennies, Nickels, and Dimes

Write how many of each coin.

	You want.	You have.	You need.
1.			   3
2.			  
3.			  
4.			  
5.			  

QUARTERS

Match.

1.     55¢

    65¢

4.

2.     50¢

5.

    90¢

3.     80¢

6.

    99¢

Problem Solving

Solve each problem.

You have this much.	You buy.	Your change.
1. 	 15¢	10 ¢
2.   	 29¢	____ ¢
3.   	 34¢	____ ¢
4.  	 35¢	____ ¢
5.   	 37¢	____ ¢
6.  	 29¢	____ ¢

Order to 100

These numbers are mixed up.
You can put them in order.

29, 17, 51, 34

17, 29, 34, 51

Try these.

1. 48, 77 38, 64

2. 11, 24, 9, 18

3. 73, 98, 19, 42

4. 12, 4, 88, 16

Connect the dots in order. Some numbers are missing.

33

NAME ______________________

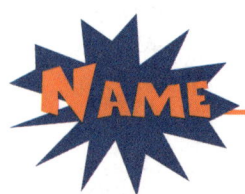

CENTIMETERS

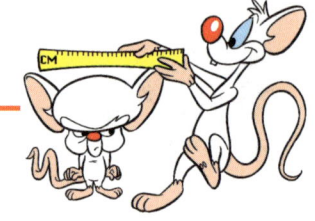

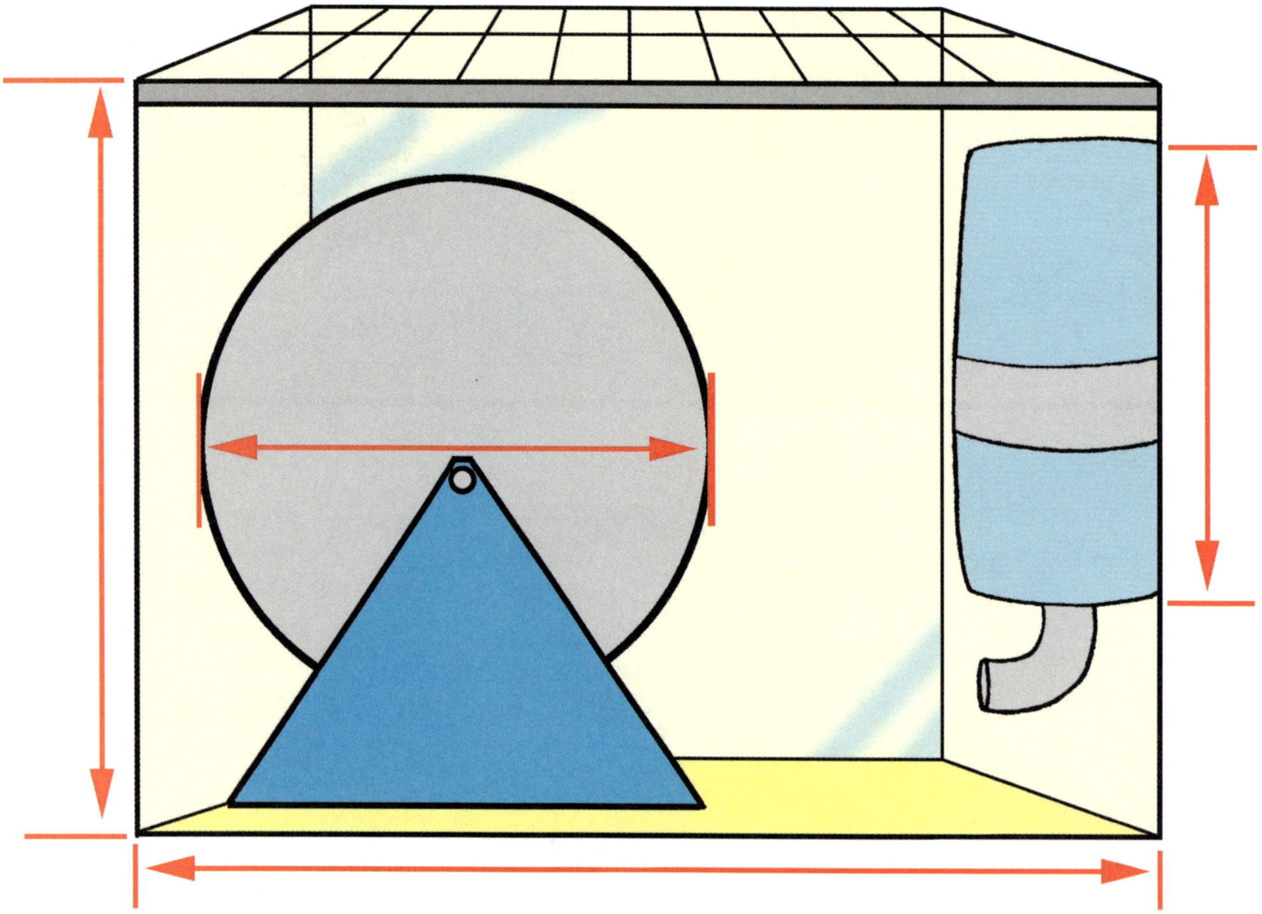

Measure the Acme Labs cage. Use a centimeter ruler.

1. The cage is about 10 cm high. ↕
2. The cage is about ______ cm wide. ↔
3. The 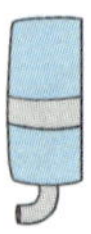 is about ______ cm high.
4. A 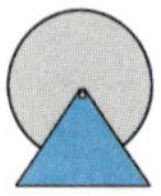 is about ______ cm wide.

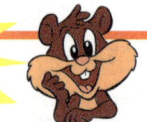

Problem Solving

Circle the set of tomatoes that will fill each place.

1.

2.

3.

4.

5.

Inch and Foot

Show which paths each bug took.

1. A → B → D = 10 inches

   I walked 10 inches

2. B → ___ → F → ___ = 11 inches

3. D → ___ → ___ → ___ = 9 inches

4. A → ___ → ___ → ___ = 1 foot

5. C → ___ → ___ → ___ → ___ = 1 foot

Problem Solving

Finish the story about the frogs.

1. 6 frogs sat on a log. 2 frogs jumped in the water.

   Then there were ___4___ frogs left on the log.

2. 3 more frogs jumped onto the log.

   Then there were _____ frogs in all.

3. 1 frog fell off the log when a fly flew by.

   Then there were _____ frogs on the log.

4. 4 more frogs climbed up on the log.

   Then there were _____ frogs on the log.

Write the number sentences for the story.

1. 6 – 2 = 4

2. _____

3. _____

4. _____

37

Name: _____

PROGRESS CHECK

Draw pictures to solve.

1. Meredith has 5 plants. Each plant has 2 flowers. How many flowers in all?

   10 flowers

2. David has 4 toy cars. Each car has 4 wheels. How many wheels in all?

   _____ wheels

3. Jon has 2 bags. Each bag has 6 marbles. How many marbles in all?

   _____ marbles

4. Jennifer has 3 bowls. Each bowl has 5 apples. How many apples in all?

   _____ apples

Facts to 11

Match the picture to the sentence. Complete.

1.
2.
3.
4.
5.
6.

3 + 8 = ___

11 − 4 = 7

6 + 5 = ___

11 − 8 = ___

11 − 5 = ___

7 + 4 = ___

Problem Solving

Read the story. Add or subtract to find the answer.

Wakko has a sweet tooth.
He loves to eat cookies.

He took 5 chocolate chip cookies and 6 peanut butter cookies from the cookie jar.

5 + 6 = 11

He gobbled 3 cookies.

Then he took 1 sugar cookie from the jar.

Wakko ate 6 cookies with a big glass of milk.

With only 3 cookies left, he searched for more. Now Wakko has a craving for raisins. So he took 7 oatmeal raisin cookies.

He gobbled 9 cookies.

Then he ate 1 more just for good measure.

How many cookies does Wakko have left?

40

More or Less

Solve.

1. Wakko bounced 10 times.
   Yakko bounced 7 times.
   Who bounced more times?
   How many more times did he bounce?

   Wakko
   3 more

2. Rita sang 11 songs.
   Slappy sang 9 songs.
   Who sang fewer songs?
   How many fewer songs did she sing?

3. Skippy jumped 6 hurdles.
   Runt jumped 8 hurdles.
   Who jumped more hurdles?
   How many more hurdles did he jump?

4. Miss Flamiel danced the hula 8 times.
   Mary Hartless danced the hula 11 times. Who danced fewer times?
   How many fewer times did she dance?

Problem Solving

A baseball team has 9 players.
Look at the teams.

SQUIRRELS   PIGEONS

1. Write a number sentence to show what each team should do to get 9 players.

   Squirrels   12 − 3 = 9

   Pigeons ___ ___ = ___

2. The Squirrels make 4 points. They make 7 more points.
   The Pigeons make 6 points. Then they make 3 more points.

Squirrels	Pigeons
9	11

   Is the scoreboard correct? _____

   Write the number sentences to show the points.

   Squirrels ___ ___ ___ = ___

   Pigeons ___ ___ ___ = ___

Circle the winner.   Squirrels   Pigeons

Facts to 12

| Names for 4: | 2 + 2 | 5 – 1 | 3 + 1 | 12 – 8 |
| Names for 5: | 3 + 2 | 6 – 1 | 4 + 1 | 12 – 7 |

Circle names for numbers.

1. 4 + 2   (2 + 5)   **2.** 5 + 4   9 – 0

8 – 1   8 + 1   3 + 5   7 + 2

5 + 3   **7**   10 – 3   6 + 3   **9**   12 – 4

11 – 5   1 + 6   1 + 8   10 – 1

4 + 3   12 – 5   11 – 2   8 + 0

3. 10 – 2   5 + 3   **4.** 12 – 6   3 + 3

4 + 4   7 + 2   10 – 6   1 + 6

12 – 4   **8**   8 – 1   9 – 3   **6**   4 + 2

8 + 0   11 – 3   0 + 5   11 – 5

5 + 2   10 – 6   10 – 4   7 + 1

PROBLEM SOLVING

Make a list to show how many combinations you can make. Color.

○ red	○ blue		□ orange	□ brown	□ yellow
△ red	△ blue		△ yellow		△ purple

○	△
red	red

_____ combinations

□	△

_____ combinations

44

Patterns

Continue each pattern.

1.
2.
3.
4.

One Half

Look at the shapes. Circle each shape that is divided into equal parts.

1.
2.
3.
4.
5.
6.

 **NAME** _____

ONE FOURTH

Show how you would share.

1. How many friends can you invite to share the pizza with you?

    friends

2. How many friends can you invite to share the sandwich?

   _____

   _ _ _ _ _ friend

3. How many friends can you invite to share the muffins?

   _____

   _ _ _ _ _ friends

4. How many friends can you invite to share the milk?

   _____

   _ _ _ _ _ friends

 NAME

ONE THIRD

Yakko, Wakko, and Dot are eating lunch. Color to show each character's share.

 RED — Yakko

 BLUE — Wakko

 YELLOW — Dot

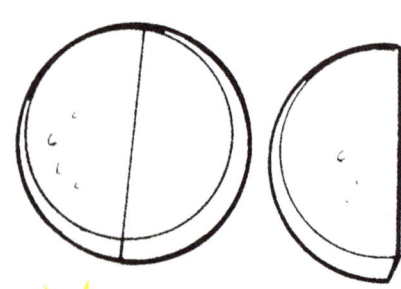

 48

Problem Solving

Solve.

1. Vince and Fred have 6 combs. They share the combs with Tom. How many combs does each person get?

   _2_ combs

   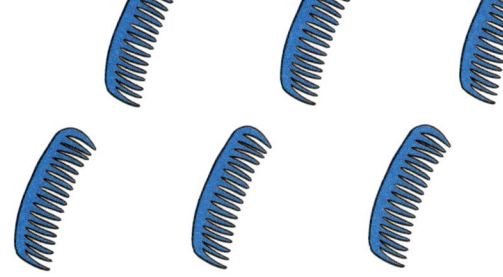

2. Pat has 4 toothbrushes. He shares them with 3 friends. How many toothbrushes does each person get?

   _____ toothbrush(es)

3. Connie has 12 tubes of toothpaste. She shares them with Kelly and Melodie. How many tubes of toothpaste does each person get?

   _____ tubes of toothpaste

4. Michelle has a bar of soap. She shares it with Rhonda. How much does each person get?

   _____ bar(s) of soap

PROGRESS CHECK

Michael writes number sentences in code. This is his code.

0	1	2	3	4	5	6	7	8	9	10	11	12
A	B	C	D	E	F	G	H	I	J	BA	BB	BC

Complete each sentence.

1. BC − D = J   J + B =
2. E + I =   H − C =
3. BB − F =   D + H =
4. G + A =   I − B =

Write the missing letters.

5. D + G = J   BA − ___ = F
6. J − ___ = J   G + ___ = BC
7. I + ___ = BA   BB − ___ = I
8. ___ + C = H   ___ + F = BB

Facts to 13

Find 3 numbers in a row that can make an addition or subtraction fact. Circle the facts. There are 13 facts in all.

(13	9	4)	7	(5	8	13)
7	4	13	11	5	6	2
8	4	12	3	2	1	4
1	10	2	8	4	3	7
3	7	6	13	2	6	5
6	8	10	6	4	3	2
12	6	6	3	7	2	9
13	6	7	5	3	2	4

Facts to 14

Color the balloons that name the number. Use a different color for each row. The first problem has been done for you.

1. 14 | (7 + 5) | (8 + 5) | **7 + 7** | **8 + 6** | **5 + 9**

2. 5 | (14 − 9) | (13 − 8) | (11 − 6) | (12 − 5) | (12 − 4)

3. 13 | (8 + 5) | (3 + 7) | (9 + 4) | (2 + 9) | (7 + 6)

4. 6 | (9 − 4) | (14 − 8) | (3 − 3) | (13 − 7) | (11 − 5)

5. 9 | (14 − 5) | (13 − 4) | (12 − 5) | (12 − 3) | (13 − 7)

6. 14 | (7 + 7) | (6 + 7) | (9 + 5) | (5 + 8) | (6 + 8)

Facts to 15

Write + or − in the ◯.

1. 12 ⎯ 6 = 6
2. 7 ◯ 6 = 13
3. 9 ◯ 5 = 14
4. 11 ◯ 6 = 5
5. 15 ◯ 8 = 7
6. 9 ◯ 0 = 9
7. 14 ◯ 5 = 9
8. 8 ◯ 8 = 0
9. 6 ◯ 8 = 14
10. 15 ◯ 9 = 6

Problem Solving

Write the price on each tag.

1. Paula spent 17¢. The 🪀 was 1¢ more than the 🎵.

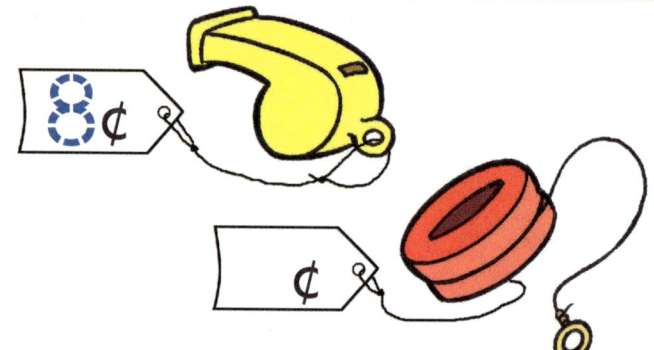

2. Beth spent 18¢. Each toy was the same price.

3. Lana spent 15¢. The 🤡 was 1¢ less than the 🐷.

4. Chris paid 1¢ more for the 🚚 than the 🚗. How much did Chris spend? _____

Problem Solving

Solve.

1. Yakko and Wakko are brothers.
   They threw a birthday party for Dot.
   Yakko asked 7 friends to come.
   Wakko asked 8 friends to come.
   How many friends did the brothers ask?

   15

2. The brothers asked _____ friends.
   4 friends could not come to the party.
   How many friends came?

3. Each person at the party will get a cupcake.
   How many cupcakes do the
   brothers have to buy?

4. Yakko poured lemonade
   for 8 friends.
   Wakko poured lemonade for the
   rest of their friends.
   How many glasses did Wakko pour?

5. 9 of their friends won prizes.
   How many friends did not
   win prizes?

TENS

Count by tens. Complete the number line.

10

Pretend you are one of the Animaniacs.
Write the number that tells where you land.

1. Start at 10.
   Hop 2 times to the right.
   Where will you land?
   30

2. Start at 50.
   Hop 3 times to the left.
   Where will you land?
   ___

3. Start at 50.
   Hop 3 times to the right.
   Where will you land?
   ___

4. Start at 70.
   Hop 6 times to the left.
   Where will you land?
   ___

5. Start at 40.
   Hop 5 times to the right.
   Where will you land?
   ___

6. Start at 60.
   Hop 3 times to the left.
   Where will you land?
   ___

7. Start at 50.
   You want to get to 10.
   What should you do?
   ___

Tens and Ones

Dot is reading a book about space. She made this chart to show how many pages she read.

A ⭐ equals 10 pages. A 🌙 equals 1 page.

NUMBER OF PAGES READ

Monday ⭐🌙	Thursday ⭐🌙🌙🌙🌙🌙🌙🌙
Tuesday 🌙🌙🌙🌙🌙🌙🌙	Friday ⭐🌙🌙
Wednesday ⭐🌙🌙🌙	Saturday ⭐🌙🌙🌙🌙🌙🌙

Use the chart to answer the questions.

1. How many pages did Dot read on Wednesday? _13_

2. On which day did Dot read 16 pages?

3. On which day did Dot read the most pages?

4. On how many days did Dot read more than 12 pages?

5. Dot read 18 pages on Sunday. How would she show this on her chart?

Adding Ones and Tens

Solve.

1. Charles had 26 stamps.
   He got 10 more.
   Does he have more than
   30 stamps? _yes_

2. Jennifer had 34 goldfish.
   She got 10 more.
   Does she have more than
   45 goldfish? _____

3. Allen had 57 books.
   He got 10 more.
   Does he have more than
   60 books? _____

4. Tracy has 32 stamps.
   She needs 42 stamps.
   How many stamps should
   she buy? _____

5. Victoria had 43 beads.
   Now she has 53 beads.
   How many more beads
   did she get? _____

Review

Complete.

ACROSS →

A. 21 + 22
C. 14 + 12
E. 34 + 53
G. 23 + 36
J. 10 + 8
L. 11 + 21
N. 34 + 15
P. 56 + 22
R. 14 + 43
T. 23 + 23

DOWN ↓

B. 15 + 23
D. 53 + 12
F. 41 + 30
H. 63 + 34
I. 21 + 42
K. 60 + 24
M. 17 + 10
O. 32 + 63
Q. 31 + 53
S. 35 + 42

59

Progress Check

Solve.

1. I am a number between 7 and 11.
   I am 2 more than 8.
   What number am I?   10

2. I am an odd number between 10 and 15.
   I am 5 more than 6.
   What number am I?   ____

3. I am a number between 1 and 10.
   I am 3 more than 7 − 3.
   What number am I?   ____

4. I am an even number between 3 and 9.
   I am 5 less than 11.
   What number am I?   ____

5. I am an odd number between 6 and 11.
   I am not the sum of 3 + 4.
   What number am I?   ____

Problem Solving

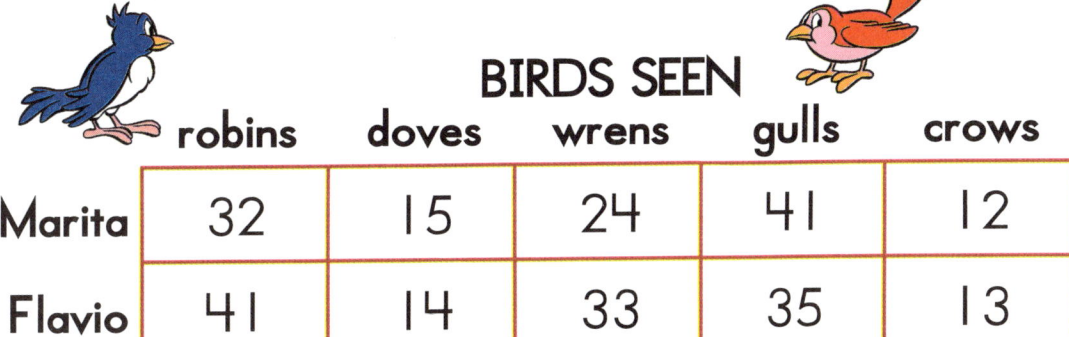

	robins	doves	wrens	gulls	crows
Marita	32	15	24	41	12
Flavio	41	14	33	35	13

Solve.

1. Who saw more wrens?   Flavio

2. Who saw more gulls?

3. How many doves did Marita and Flavio see in all?

4. How many robins and crows did Marita see?

5. How many doves and robins did Flavio see?

6. Did Flavio see more wrens than gulls?

7. Did Marita see more crows, doves, or robins?

Subtracting Ones and Tens

Subtract each number in the middle circle from the number in the center to complete each wheel.

1.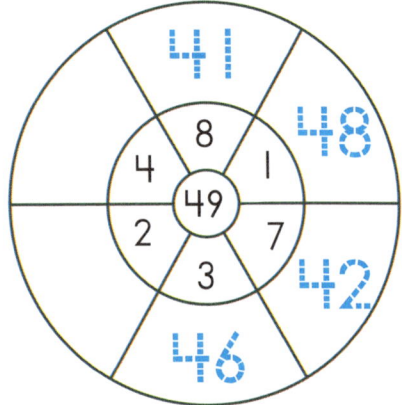

2.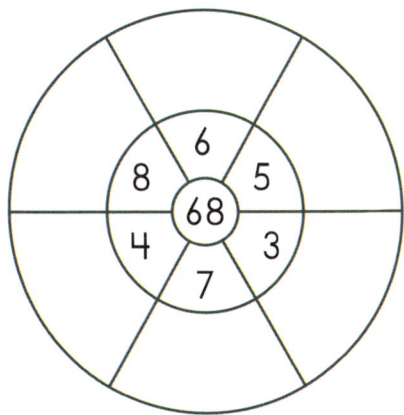

3.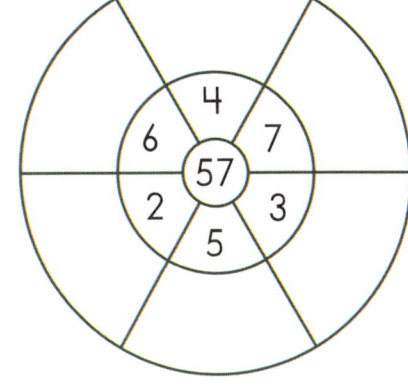

4.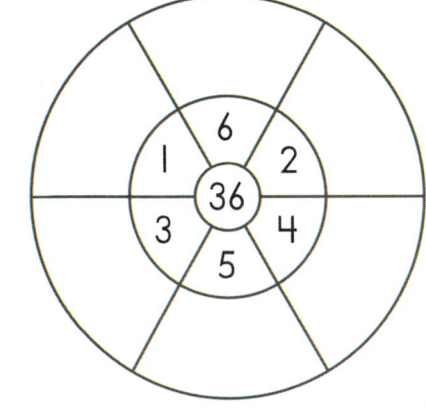

Progress Check

Subtract.
Draw a smile if there is 10¢ or more left.
Draw a frown if there is less than 10¢.

99¢	99¢	99¢	99¢
− 13¢	− 11¢	− 14¢	− 12¢
86¢			
− 21¢	− 23¢	− 10¢	− 31¢
65¢			
− 11¢	− 20¢	− 42¢	− 12¢
54¢			
− 32¢	− 23¢	− 23¢	− 10¢
− 11¢	− 20¢	− 3¢	− 24¢

Estimating

Circle the best estimate.

1. Morgan saw 11 lions at the zoo. She also saw 12 tigers and 19 leopards. About how many of these animals did she see?

   20 animals

   (40 animals)

2. Spencer saw 21 pink birds dance. He saw 10 parrots ride bicycles. Then Spencer saw 32 white birds singing. About how many birds did he see?

   60 birds

   70 birds

3. Brooke counted 29 blue fish. She also counted 52 grey fish and 13 yellow fish. About how many fish did she count?

   70 fish

   90 fish

4. Cody watched 29 white ducks and 22 brown ducks swimming. Then 20 ducks flew away. About how many ducks were left?

   30 ducks

   50 ducks

64

Answer Key

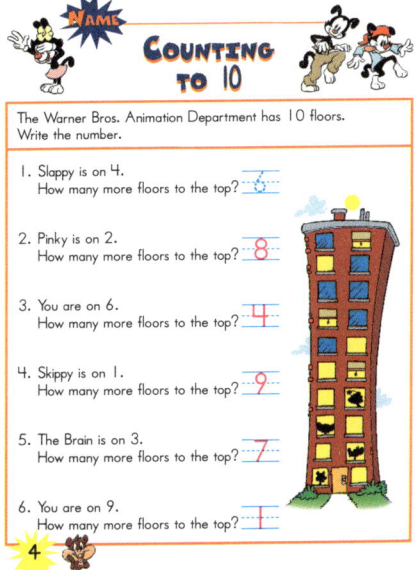
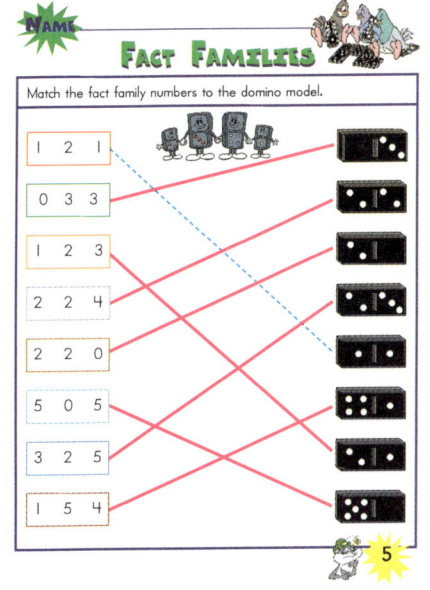

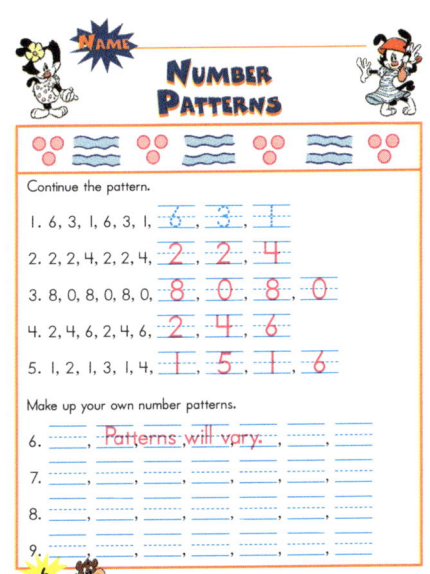

65

Answer Key

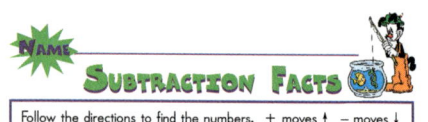
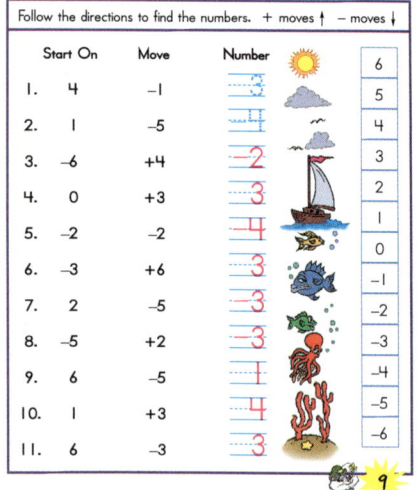
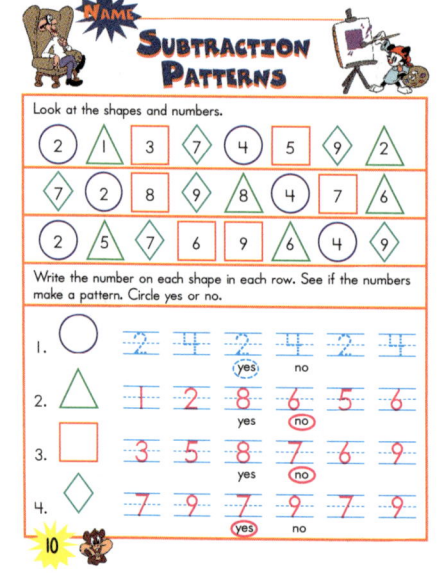
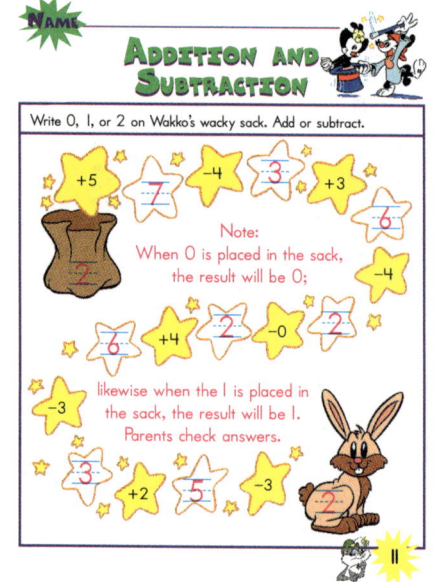

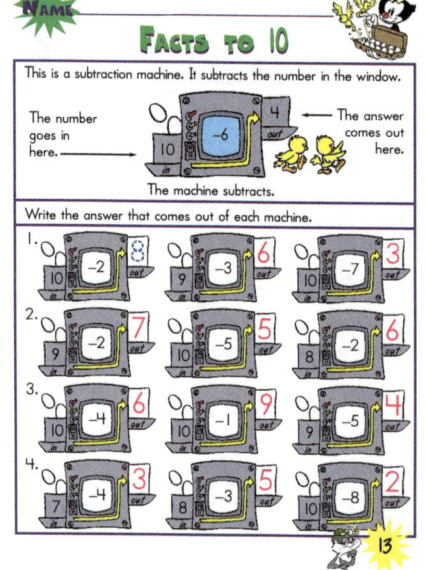

Answer Key

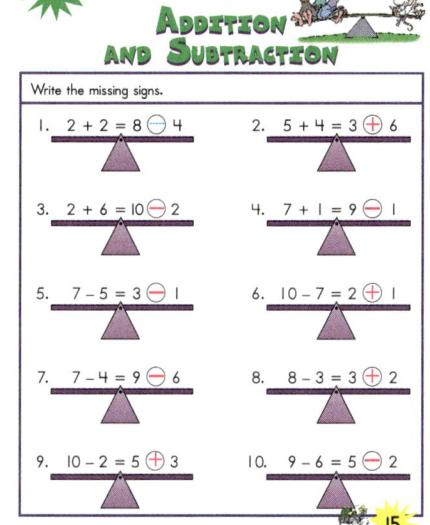
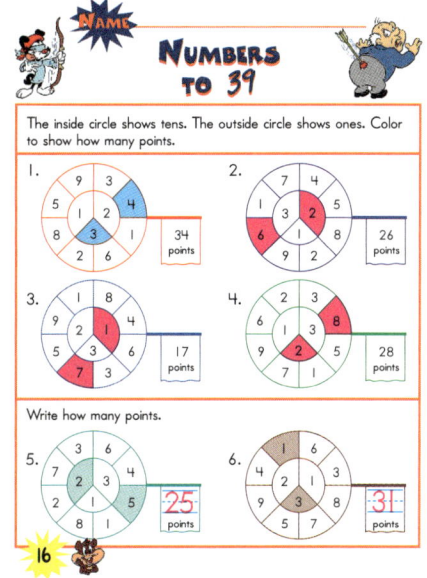
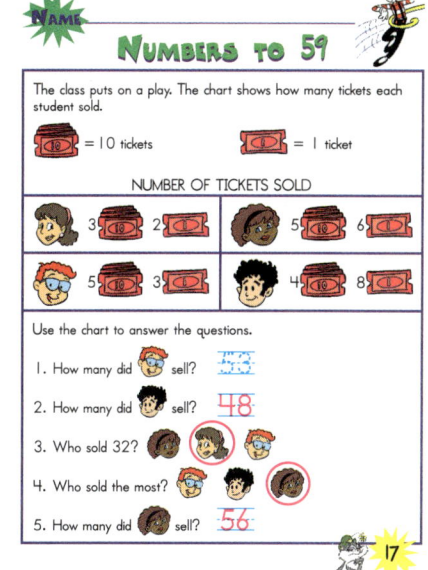
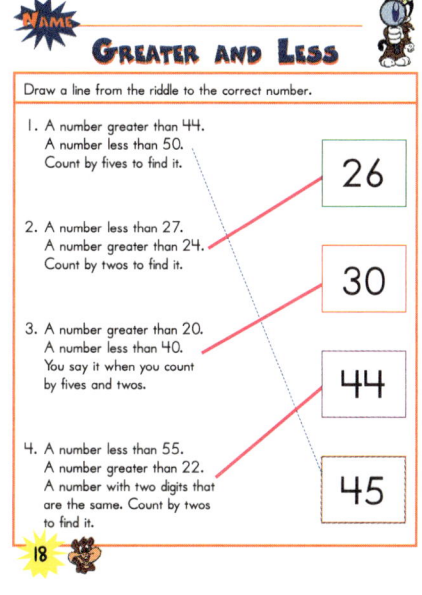
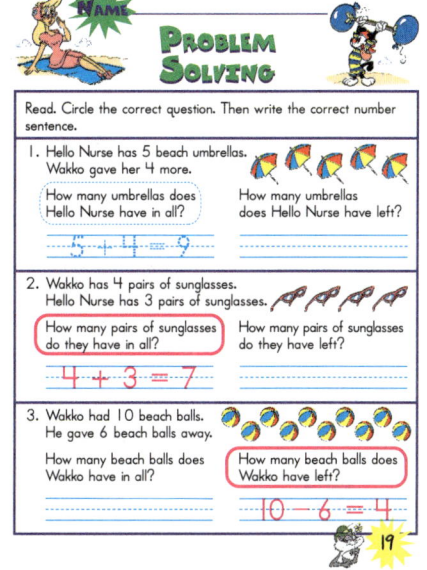

67

Answer Key

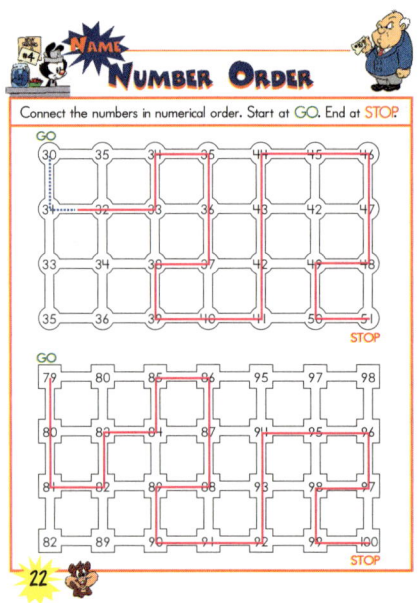
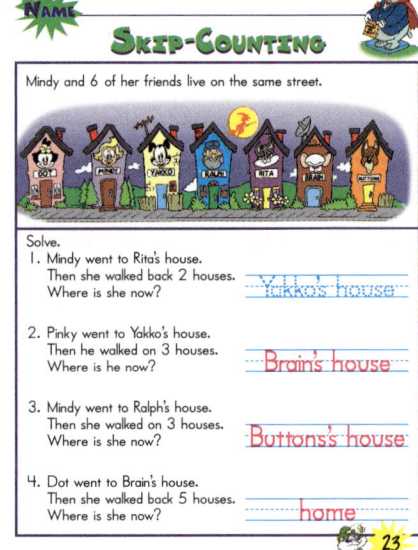
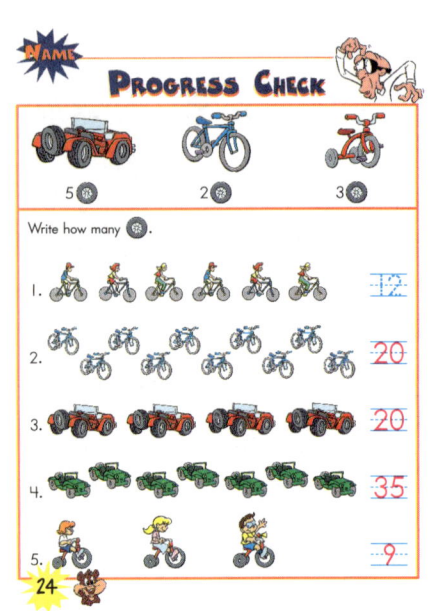
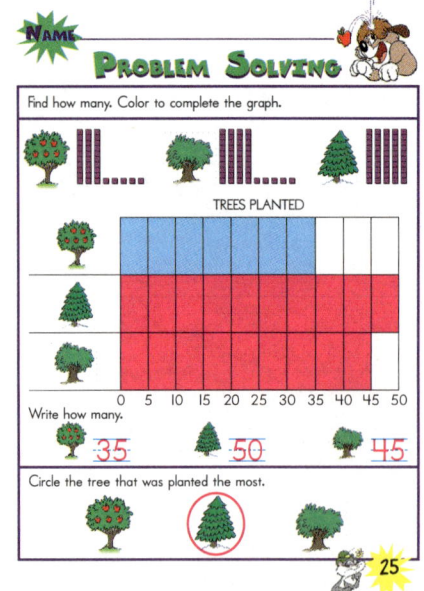

Answer Key

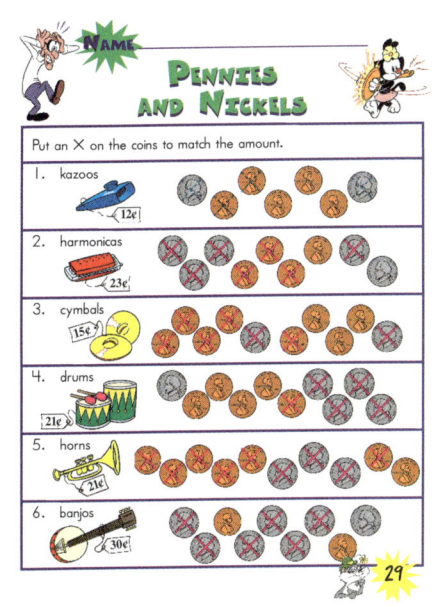
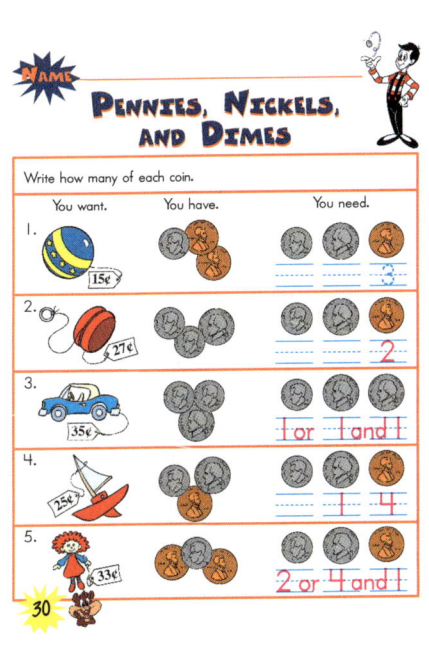
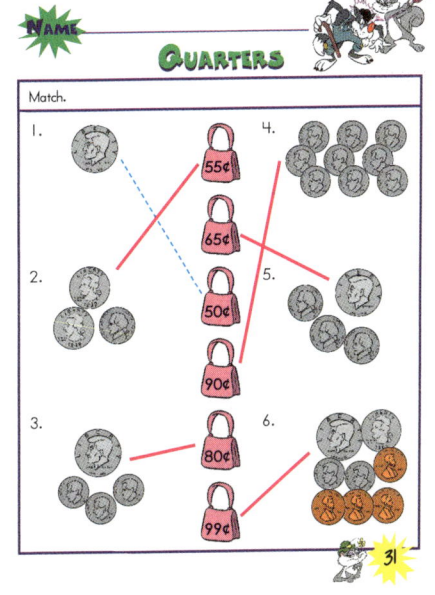

Answer Key

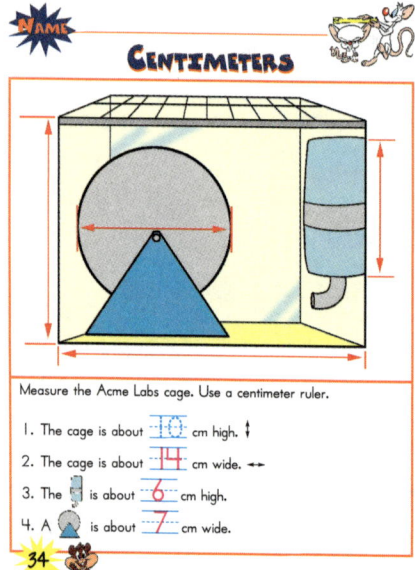
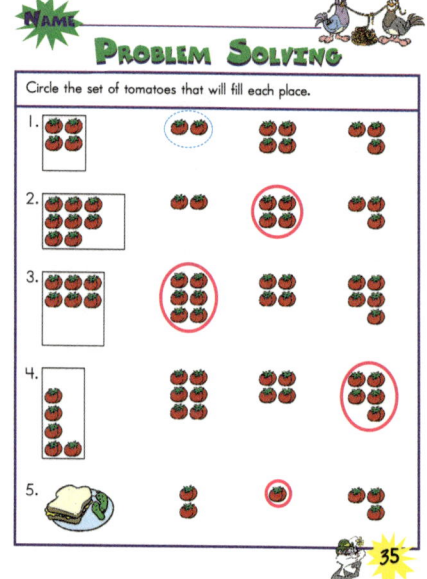
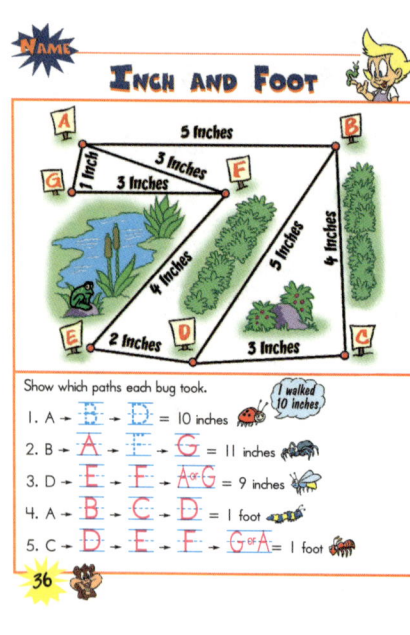

ANSWER KEY

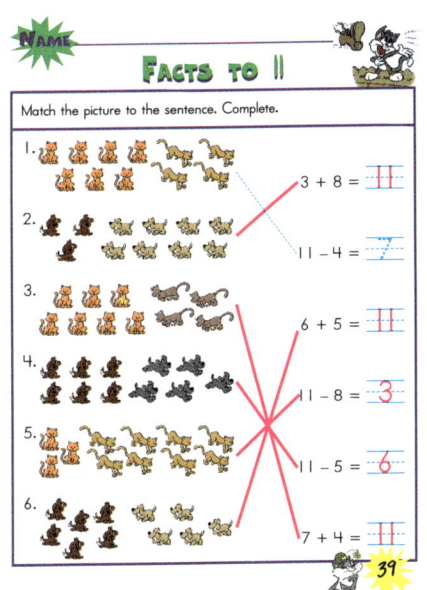
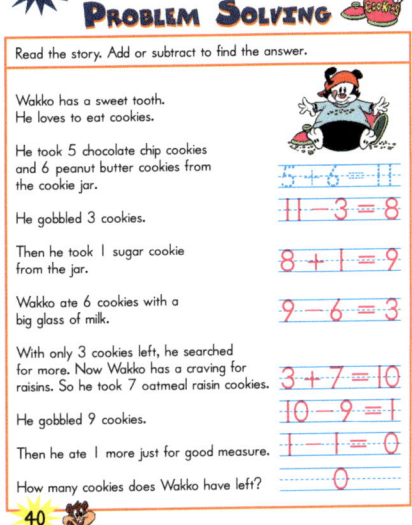
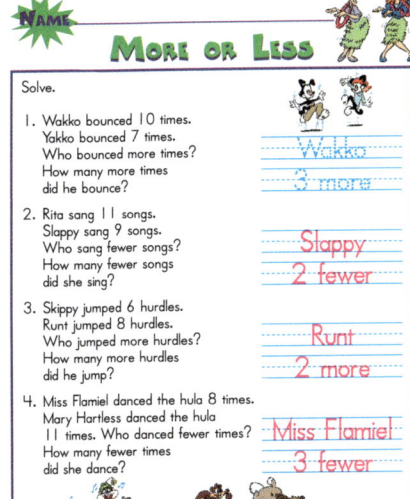
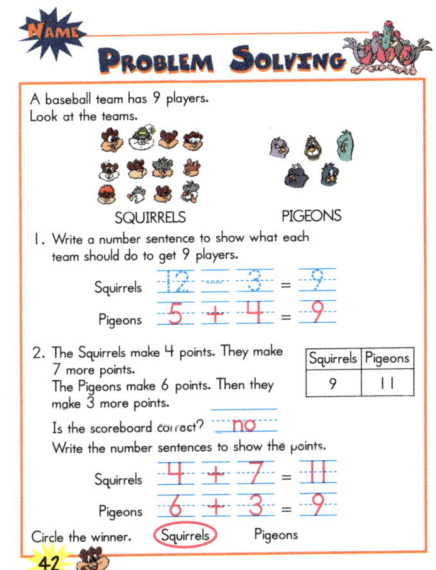
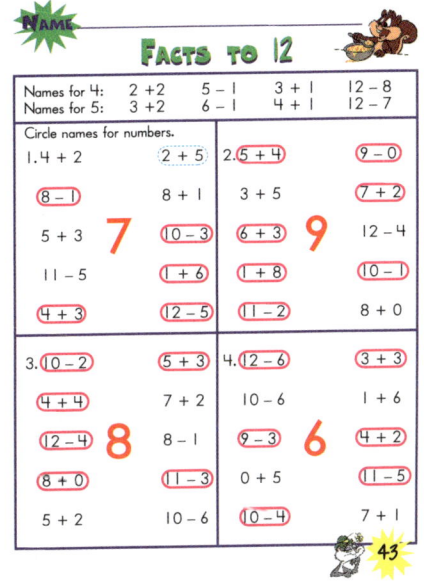
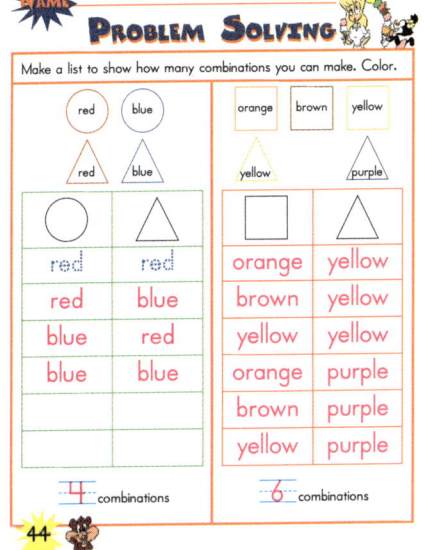

Answer Key

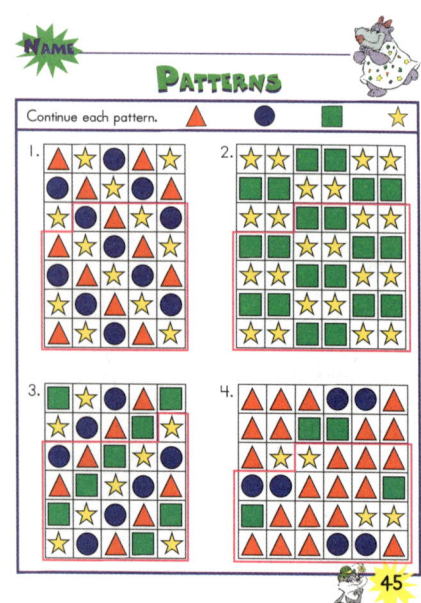
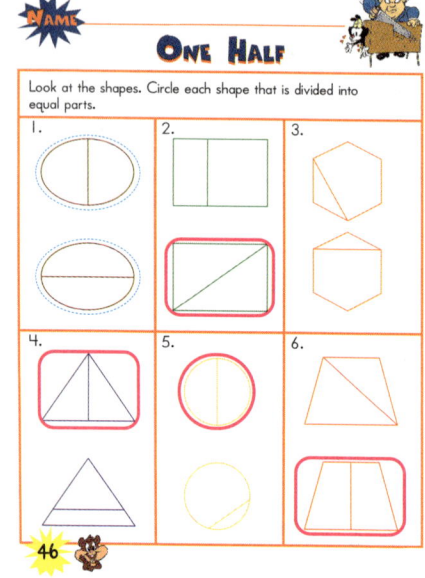
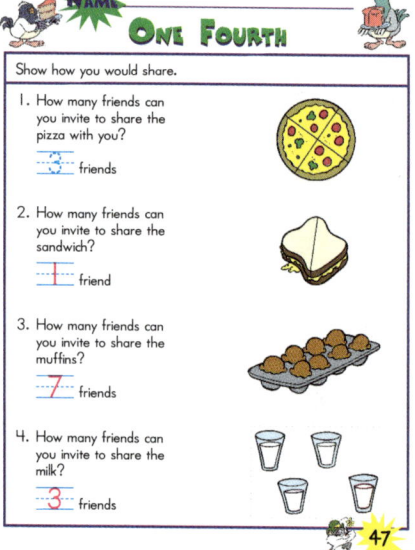

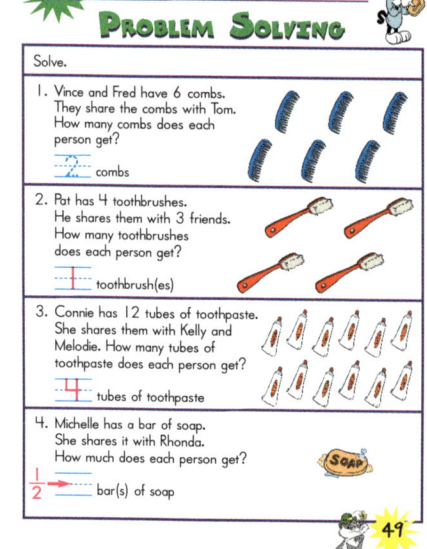
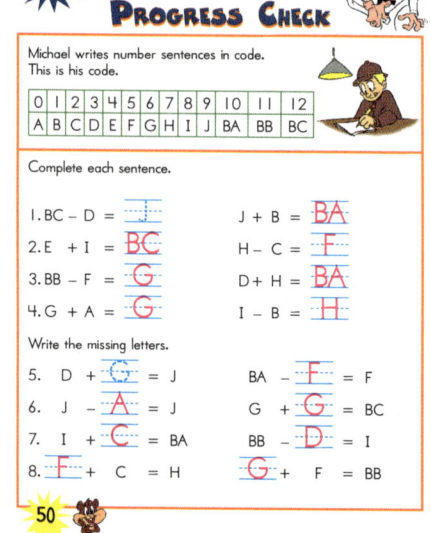

72

Answer Key

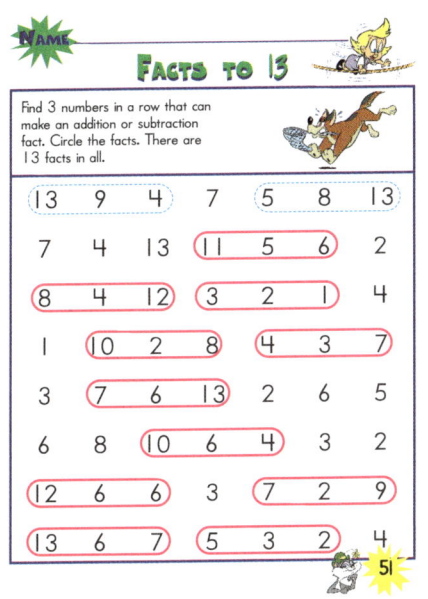
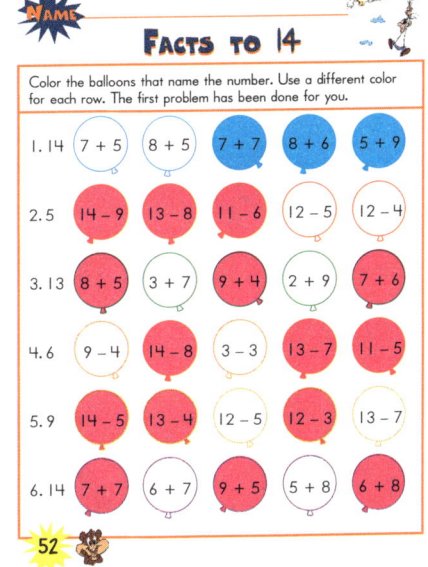

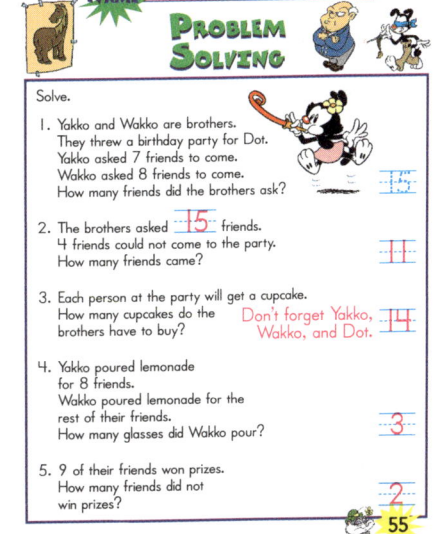

Answer Key

Tens and Ones (57)

Dot is reading a book about space. She made this chart to show how many pages she read.

A ★ equals 10 pages. A ☾ equals 1 page.

NUMBER OF PAGES READ

Monday	★ ☾	Thursday	★ ☾☾☾☾☾
Tuesday	☾☾☾☾☾	Friday	★ ☾
Wednesday	★ ☾☾☾	Saturday	★ ☾☾☾☾☾☾

Use the chart to answer the questions.

1. How many pages did Dot read on Wednesday? **13**
2. On which day did Dot read 16 pages? **Saturday**
3. On which day did Dot read the most pages? **Thursday**
4. On how many days did Dot read more than 12 pages? **3**
5. Dot read 18 pages on Sunday. How would she show this on her chart? **★☾☾☾☾☾☾☾☾**

Adding Ones and Tens (58)

Solve.

1. Charles had 26 stamps. He got 10 more. Does he have more than 30 stamps? **yes**
2. Jennifer had 34 goldfish. She got 10 more. Does she have more than 45 goldfish? **no**
3. Allen had 57 books. He got 10 more. Does he have more than 60 books? **yes**
4. Tracy has 32 stamps. She needs 42 stamps. How many stamps should she buy? **10**
5. Victoria had 43 beads. Now she has 53 beads. How many more beads did she get? **10**

Review (59)

Crossword puzzle answers:
Row 1: 4, 3, 2, 6
Row 2: 8, 7, 5, 9
Row 3: 6, 1, 8, 7
Row 4: 3, 2, 9, 5
Row 5: 7, 8, 5, 7
Row 6: 4, 6, 7

Complete.

ACROSS →
A. 21 + 22
C. 14 + 12
E. 34 + 53
G. 23 + 36
J. 10 + 8
L. 11 + 21
N. 34 + 15
P. 56 + 22
R. 14 + 43
T. 23 + 23

DOWN ↓
B. 15 + 23
D. 53 + 12
F. 41 + 30
H. 63 + 34
I. 21 + 42
K. 60 + 24
M. 17 + 10
O. 32 + 63
Q. 31 + 53
S. 35 + 42

Progress Check (60)

Solve.

1. I am a number between 7 and 11. I am 2 more than 8. What number am I? **10**
2. I am an odd number between 10 and 15. I am 5 more than 6. What number am I? **11**
3. I am a number between 1 and 10. I am 3 more than 7 − 3. What number am I? **7**
4. I am an even number between 3 and 9. I am 5 less than 11. What number am I? **6**
5. I am an odd number between 6 and 11. I am not the sum of 3 + 4. What number am I? **9**

Problem Solving (61)

BIRDS SEEN

	robins	doves	wrens	gulls	crows
Marita	32	15	24	41	12
Flavio	41	14	33	35	13

Solve.

1. Who saw more wrens? **Flavio**
2. Who saw more gulls? **Marita**
3. How many doves did Marita and Flavio see in all? **29**
4. How many robins and crows did Marita see? **44**
5. How many doves and robins did Flavio see? **55**
6. Did Flavio see more wrens than gulls? **no**
7. Did Marita see more crows, doves, or robins? **robins**

Subtracting Ones and Tens (62)

Subtract each number in the middle circle from the number in the center to complete each wheel.

1. Center 49, middle ring 8,4,7,3: **45, 41, 48, 47, 42, 46**
2. Center 68, middle ring 6,8,5,3: **60, 62, 63, 65, 61, 64**
3. Center 57, middle ring 6,4,7,2: **51, 53, 50, 54, 52, 55**
4. Center 36, middle ring 6,1,2,3: **35, 30, 34, 32, 31, 33**

74

Answer Key

Progress Check

Subtract.
Draw a smile if there is 10¢ or more left.
Draw a frown if there is less than 10¢.

99¢	99¢	99¢	99¢
− 13¢	− 11¢	− 14¢	− 12¢
86¢	88¢	85¢	87¢
− 21¢	− 23¢	− 10¢	− 31¢
65¢	65¢	75¢	56¢
− 11¢	− 20¢	− 42¢	− 12¢
54¢	45¢	33¢	44¢
− 32¢	− 23¢	− 23¢	− 10¢
22¢	22¢	10¢	34¢
− 11¢	− 20¢	− 3¢	− 24¢
11¢	2¢	7¢	10¢

63

Estimating

Circle the best estimate.

1. Morgan saw 11 lions at the zoo. She also saw 12 tigers and 19 leopards. About how many of these animals did she see?
   20 animals
   (40 animals)

2. Spencer saw 21 pink birds dance. He saw 10 parrots ride bicycles. Then Spencer saw 32 white birds singing. About how many birds did he see?
   (60 birds)
   70 birds

3. Brooke counted 29 blue fish. She also counted 52 grey fish and 13 yellow fish. About how many fish did she count?
   70 fish
   (90 fish)

4. Cody watched 29 white ducks and 22 brown ducks swimming. Then 20 ducks flew away. About how many ducks were left?
   (30 ducks)
   50 ducks

64

McGraw-Hill Consumer Products

The skills taught in school are now available at home! These award-winning software titles meet school guidelines and are based on The McGraw-Hill Companies classroom software titles.

MATH GRADES 1 & 2

These math programs are a great way to teach and reinforce skills used in everyday situations. Fun, friendly characters need help with their math skills. Everyone's friend, Nubby the stubby pencil, will help kids master the math in the Numbers Quiz show. Foggy McHammer, a carpenter, needs some help building his playhouse so that all the boards will fit together! Julio Bambino's kitchen antics will surely burn his pastries if you don't help him set the clock timer correctly! We can't forget Turbo Tomato, a fruit with a passion for adventure, who needs help calculating his daredevil stunts.

Math Grades 1 & 2 use a tested, proven approach to reinforcing your child's math skills while keeping him or her intrigued with Nubby and his collection of crazy friends.

TITLE
Grade 1: Nubby's Quiz Show
Grade 2: Foggy McHammer's Treehouse

MISSION MASTERS™ MATH AND LANGUAGE ARTS

The Mission Masters™—Pauline, Rakeem, Mia, and T.J.—need your help. The Mission Masters™ are a team of young agents working for the Intelliforce Agency, a high-level cooperative whose goal is to maintain order on our rather unruly planet. From within the agency's top secret Command Control Center, the agency's central computer, M5, has detected a threat…and guess what—you're the agent assigned to the mission!

MISSION MASTERS™ MATH GRADES 3, 4, & 5

This series of exciting activities encourages young mathematicians to challenge themselves and their math skills to overcome the perils of villains and other planetary threats. Skills reinforced include: analyzing and solving real-world problems, estimation, measurements, geometry, whole numbers, fractions, graphs, and patterns.

TITLE
Grade 3: Mission Masters™ Defeat Dirty D!
Grade 4: Mission Masters™ Alien Encounter
Grade 5: Mission Masters™ Meet Mudflat Moe

MISSION MASTERS™ LANGUAGE ARTS GRADES 3, 4, & 5

This series invites children to apply their language skills to defeat unscrupulous characters and to overcome other earthly dangers. Skills reinforced include: language mechanics and usage, punctuation, spelling, vocabulary, reading comprehension, and creative writing.

TITLE
Grade 3: Mission Masters™ Freezing Frenzy
Grade 4: Mission Masters™ Network Nightmare
Grade 5: Mission Masters™ Mummy Mysteries

BASIC SKILLS BUILDER K to 2 – THE MAGIC APPLEHOUSE

At the Magic Applehouse, children discover that Abigail Appleseed runs a deliciously successful business selling apple pies, tarts, and other apple treats. Enthusiasm grows as children join in the fun of helping Abigail run her business. Along the way they'll develop computer and entrepreneurial skills to last a lifetime. They will run their own business – all while they're having bushels of fun!

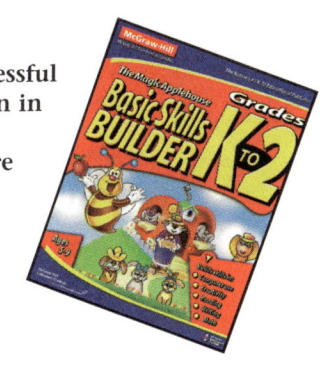

TITLE
Basic Skills Builder –The Magic Applehouse

TEST PREP – SCORING HIGH

This grade-based testing software will help prepare your child for standardized achievement tests given by his or her school. Scoring High specifically targets the skills required for success on the Stanford Achievement Test (SAT) for grades three through eight. Lessons and test questions follow the same format and cover the same content areas as questions appearing on the actual SAT tests. The practice tests are modeled after the SAT test-taking experience with similar directions, number of questions per section, and bubble-sheet answer choices.

Scoring High is a child's first-class ticket to a winning score on standardized achievement tests!

TITLE
Grades 3 to 5: Scoring High Test Prep
Grades 6 to 8: Scoring High Test Prep

SCIENCE

Mastering the principles of both physical and life science has never been so FUN for kids grades six and above as it is while they are exploring McGraw-Hill's edutainment software!

TITLE
Grades 6 & up: Life Science
Grades 8 & up: Physical Science

REFERENCE

The National Museum of Women in the Arts has teamed with McGraw-Hill Consumer Products to bring you this superb collection available for your enjoyment on CD-ROM.

This special collection is a visual diary of 200 women artists from the Renaissance to the present, spanning 500 years of creativity.

You will discover the art of women who excelled in all the great art movements of history. Artists who pushed the boundaries of abstract, genre, landscape, narrative, portrait, and still-life styles; as well as artists forced to push the societal limits placed on women through the ages.

TITLE
Women in the Arts

Most titles for Windows 3.1™, Windows '95™ & '98™, and Macintosh™.

Visit us on the Internet at:

www.MHkids.com

Or call 800-298-4119 for your local retailer.

McGraw-Hill Consumer Products

All our workbooks meet school curriculum guidelines and correspond to The McGraw-Hill Companies classroom textbooks.

SPECTRUM SERIES

DOLCH Sight Word Activities
The DOLCH Sight Word Activities Workbooks use the classic Dolch list of 220 basic vocabulary words that make up from 50% to 75% of all reading matter that children ordinarily encounter. Since these words are ordinarily recognized on sight, they are called *sight words*. Volume 1 includes 110 sight words. Volume 2 covers the remainder of the list. Over 160 pages.

TITLE	ISBN	PRICE
Grades K-1 Vol. 1	1-57768-429-X	$9.95
Grades K-1 Vol. 2	1-57768-439-7	$9.95

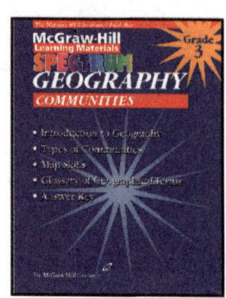
GEOGRAPHY
Full-color, three-part lessons strengthen geography knowledge and map reading skills. Focusing on five geographic themes including location, place, human/environmental interaction, movement, and regions. Over 150 pages. Glossary of geographical terms and answer key included.

TITLE	ISBN	PRICE
Gr 3, Communities	1-57768-153-3	$7.95
Gr 4, Regions	1-57768-154-1	$7.95
Gr 5, USA	1-57768-155-X	$7.95
Gr 6, World	1-57768-156-8	$7.95

MATH
Features easy-to-follow instructions that give students a clear path to success. This series has comprehensive coverage of the basic skills, helping children to master math fundamentals. Over 150 pages. Answer key included.

TITLE	ISBN	PRICE
Grade 1	1-57768-111-8	$6.95
Grade 2	1-57768-112-6	$6.95
Grade 3	1-57768-113-4	$6.95
Grade 4	1-57768-114-2	$6.95
Grade 5	1-57768-115-0	$6.95
Grade 6	1-57768-116-9	$6.95
Grade 7	1-57768-117-7	$6.95
Grade 8	1-57768-118-5	$6.95

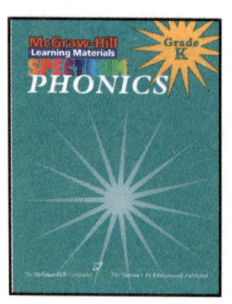
PHONICS
Provides everything children need to build multiple skills in language. Focusing on phonics, structural analysis, and dictionary skills, this series also offers creative ideas for using phonics and word study skills in other language arts. Over 200 pages. Answer key included.

TITLE	ISBN	PRICE
Grade K	1-57768-120-7	$6.95
Grade 1	1-57768-121-5	$6.95
Grade 2	1-57768-122-3	$6.95
Grade 3	1-57768-123-1	$6.95
Grade 4	1-57768-124-X	$6.95
Grade 5	1-57768-125-8	$6.95
Grade 6	1-57768-126-6	$6.95

SPECTRUM SERIES – continued

READING
This full-color series creates an enjoyable reading environment, even for below-average readers. Each book contains captivating content, colorful characters, and compelling illustrations, so children are eager to find out what happens next. Over 150 pages. Answer key included.

TITLE	ISBN	PRICE
Grade K	1-57768-130-4	$6.95
Grade 1	1-57768-131-2	$6.95
Grade 2	1-57768-132-0	$6.95
Grade 3	1-57768-133-9	$6.95
Grade 4	1-57768-134-7	$6.95
Grade 5	1-57768-135-5	$6.95
Grade 6	1-57768-136-3	$6.95

SPELLING
This full-color series links spelling to reading and writing and increases skills in words and meanings, consonant and vowel spellings, and proofreading practice. Over 200 pages. Speller dictionary and answer key included.

TITLE	ISBN	PRICE
Grade 1	1-57768-161-4	$7.95
Grade 2	1-57768-162-2	$7.95
Grade 3	1-57768-163-0	$7.95
Grade 4	1-57768-164-9	$7.95
Grade 5	1-57768-165-7	$7.95
Grade 6	1-57768-166-5	$7.95

WRITING
Lessons focus on creative and expository writing using clearly stated objectives and pre-writing exercises. Eight essential reading skills are applied. Activities include main idea, sequence, comparison, detail, fact and opinion, cause and effect, and making a point. Over 130 pages. Answer key included.

TITLE	ISBN	PRICE
Grade 1	1-57768-141-X	$6.95
Grade 2	1-57768-142-8	$6.95
Grade 3	1-57768-143-6	$6.95
Grade 4	1-57768-144-4	$6.95
Grade 5	1-57768-145-2	$6.95
Grade 6	1-57768-146-0	$6.95
Grade 7	1-57768-147-9	$6.95
Grade 8	1-57768-148-7	$6.95

TEST PREP
From the Nation's #1 Testing Company
Prepares children to do their best on current editions of the five major standardized tests. Activities reinforce test-taking skills through examples, tips, practice, and timed exercises. Subjects include reading, math, and language. Over 150 pages. Answer key included.

TITLE	ISBN	PRICE
Grade 1	1-57768-101-0	$8.95
Grade 2	1-57768-102-9	$8.95
Grade 3	1-57768-103-7	$8.95
Grade 4	1-57768-104-5	$8.95
Grade 5	1-57768-105-3	$8.95
Grade 6	1-57768-106-1	$8.95
Grade 7	1-57768-107-X	$8.95
Grade 8	1-57768-108-8	$8.95

Visit us on the Internet at:
www.MHkids.com

A McGraw・Hill/Warner Bros. Workbook

CERTIFICATE OF ACCOMPLISHMENT

THIS CERTIFIES THAT

_____

HAS SUCCESSFULLY COMPLETED
THE JUNIOR ACADEMIC'S™

Grade 1 Enrichment Math
WORKBOOK.
CONGRATULATIONS AND THAT'S ALL FOLKS!

The McGraw・Hill Companies
PUBLISHER

Bugs Bunny
BUGS BUNNY, EDITOR-IN-CHIEF

RECEIVE THE McGRAW-HILL PARENT NEWSLETTER

FREE!

Thank you for expressing interest in the successful education of your child. With the purchase of this workbook, we know that you are committed to your child's development and future success. We at **McGraw-Hill Consumer Products** would like to help you make a difference in the education of your child by offering a quarterly newsletter that provides current topics on education and activities that you and your child can work on together.

To receive a free copy of our newsletter, please provide us with the following information:

Name _____

Address _____

City _____ State ____ Zip ____

e-mail (if applicable) _____

Store where book purchased _____

Grade Level of book purchased _____

Title of book purchased _____

Mail to:
Parent Newsletter
c/o McGraw-Hill Consumer Products
251 Jefferson Street, M.S. #12
Waldoboro, ME 04572

Or Call 800-298-4119

Or visit us at:

www.MHkids.com

This offer is limited to residents of the United States and Canada and is only in effect for as long as the newsletter is published.
The information that you provide will not be given, rented, or sold to any company.